W0079587

The World of Energy

Raj Sharma · Vishnu Pareek

The World of Energy

Engine of Life

Understanding that there is no free lunch—
caught between a rock and a hard place!

 Springer

Raj Sharma
Curtin University
Perth, WA, Australia

Vishnu Pareek
Curtin University
Perth, WA, Australia

ISBN 978-981-15-6723-0   ISBN 978-981-15-6724-7   (eBook)
https://doi.org/10.1007/978-981-15-6724-7

© Springer Nature Singapore Pte Ltd. 2020, corrected publication 2021
This work is subject to copyright. All rights are reserved by the Publisher, whether the whole or part of the material is concerned, specifically the rights of translation, reprinting, reuse of illustrations, recitation, broadcasting, reproduction on microfilms or in any other physical way, and transmission or information storage and retrieval, electronic adaptation, computer software, or by similar or dissimilar methodology now known or hereafter developed.
The use of general descriptive names, registered names, trademarks, service marks, etc. in this publication does not imply, even in the absence of a specific statement, that such names are exempt from the relevant protective laws and regulations and therefore free for general use.
The publisher, the authors and the editors are safe to assume that the advice and information in this book are believed to be true and accurate at the date of publication. Neither the publisher nor the authors or the editors give a warranty, expressed or implied, with respect to the material contained herein or for any errors or omissions that may have been made. The publisher remains neutral with regard to jurisdictional claims in published maps and institutional affiliations.

Cover credits: Garry Killian/shutterstock

This Springer imprint is published by the registered company Springer Nature Singapore Pte Ltd.
The registered company address is: 152 Beach Road, #21-01/04 Gateway East, Singapore 189721, Singapore

It is for us to understand the Laws of Nature—not for them to explain themselves!

Dedicated to Our Parents

Mrs. Shanta Sharma and Late Dr. Shyam Lal Sharma

—Raj Sharma

Mrs. Santosh Pareek and Late Mr. Madan Lal Pareek

—Vishnu Pareek

Preface

The Arab Oil embargo of the early 1970s created a panic reaction in the West. Fearing the end of oil, people formed miles long queues at gas stations (petrol pumps) waiting to fill a tank of gas (gasoline; petrol). Calls for alternate energy began, which waxed and waned with fluctuating oil prices over the next couple of decades even though alternate energy and oil do not have much in common. Calls for alternate energy have metamorphosed over the past few decades into calls for renewable energy, clean energy and now green energy with global warming and climate change as raison d'etre, pinning it on fossil fuel use. Perhaps, this is what has driven increased media coverage of matters-energy in recent years.

Energy drives all life. Without energy, there is no life. Adapting a common saying, money alone cannot make the world go around, it is energy that does! In view of the renewed interest in energy among public at large, including policy makers, who are too preoccupied to gain a formal understanding of the field of energy via attending lectures or from textbooks, this slim book attempts to provide a sound introduction to the field of energy. What is energy? Why is it important? Where and how did it all begin? Role it has played in the development of modern life? Current status? The alternatives and limitations? Nature's laws governing energy? Should the reader gain an understanding of the fundamentals of energy to arrive at informed conclusions, we would have succeeded. Perhaps, this will also spark an informed debate on the way forward.

Perth, Australia   Raj Sharma
                                                                                             Vishnu Pareek

Acknowledgements

We gratefully acknowledge the helpful inputs and feedback of (not in any order) Dr. Shashi Panicker, Dr. Nilanjan Chattopadhyay, Mr. Payonidhi Kasliwal, Dr. Javed Iqbal, Dr. Awdhesh Kumar, Mr. Rishabh Kasliwal and Dr. Ganesh N. Saxena. We also acknowledge the initial discussions with Dr. Gaurav Bhaduri and Dr. Harsh Pandey before the manuscript was written. *Thank you!*

We are deeply indebted to Rati and Neeraj (daughter and son-in-law of Raj Sharma) for critiquing the many versions of the transcript as it was being written. Especially Rati, whose constant encouragement, support and inputs saw the culmination of this effort. *Thank you!*

Finally, we say *'thank you'* to the scores of our students over the past few decades who contributed directly or indirectly to our understanding of 'energy' resulting in this book.

Raj Sharma
Vishnu Pareek

About the Book

Energy is the capacity to do work—from the mundane to the complex. Energy drives life. Solar energy, in its existing form and intensity, sustains earth—take the Sun away, and all life on earth, as we know it, shall cease to exist. Sun is responsible for all natural energy—including fossil fuels, wind, biomass, others. Laws governing energy are laws of nature—eternal and inviolable. All energy cannot be converted in to work—not only can you not win, you cannot breakeven either; this is the law of nature. Evolution of energy use, by amount and type, over the past couple of hundred years, and its importance on the development of modern life, are presented. Different forms of energy are discussed and questions raised on the sustainability of the present-day model of development. Climate change is real but there is a quick peek in to whether it could be man-made or not.

The only and inviolable truth about energy is that all energy, alternate energy included, is governed by, and must obey, certain laws of Nature, called Laws of Thermodynamics. There is a cost to everything and a price has to be paid.

There is no free lunch—caught between a rock and a hard place!

Contents

Chapter 1
Introduction—*Setting the Stage*

1.1 Teeing-Up

Energy has been front and center on the world stage ever since the Arab Oil Embargo on the West in the early 1970s following the Arab—Israeli Yom Kippur War. The oil embargo threw the West into panic, which was used to cheap oil (Fig. 1.1), with mile long queues at gas stations (petrol pumps) to fill up a tank of gasoline (petrol) as if the end of oil was near. Crude oil prices almost quadrupled overnight—the days of cheap oil were over. Middle-Eastern countries, led by Saudi Arabia (in 1974; Iran had already nationalized its oil in 1951), began nationalizing oil companies and regaining control over their natural resource long exploited by the American and British multinational oil companies. OPEC (Organization of Petroleum Exporting Countries), lying dormant ever since its founding in Iraq in 1960, had suddenly become strong!

Gradually, the world came to order again and settled to a new normal—higher oil prices and OPEC influence.

Industrialization and development of the modern way of life in the West happened on the back of cheap, reliable, and abundant energy—coal, oil and natural gas. The oil embargo, and its nationalization by the Arab countries, led to cries for 'alternate energy' (solar/wind/geothermal/others) in the West and a reduced dependence on oil (world is consuming more oil today than it was in the 1970s!). There was (and is) a major fallacy in this—'alternate energy', unless used directly as in the past over millennia, delivers but one product 'electricity'—and 'electricity' and 'oil' do not 'mix', with oil accounting for only about 6% of the world's electricity generation even today and most of it being generated using coal and natural gas.

President Jimmy Carter, the 39th President of the United States, announced the formation of a new Department of Energy in the mid-1970s and setting up of the Solar Energy Research Institute (SERI) in Golden, Colorado (SERI has seen many ups and downs over the past four decades with changing US administrations and is now known as 'National Renewable Energy Lab' (NREL)).

© Springer Nature Singapore Pte Ltd. 2020
R. Sharma and V. Pareek, *The World of Energy*,
https://doi.org/10.1007/978-981-15-6724-7_1

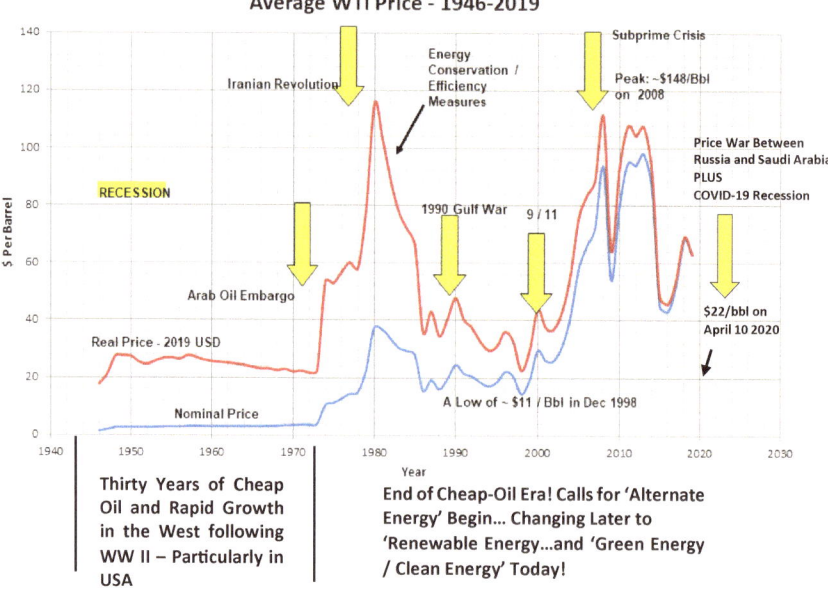

Thirty Years of Cheap Oil and Rapid Growth in the West following WW II – Particularly in USA

End of Cheap-Oil Era! Calls for 'Alternate Energy' Begin... Changing Later to 'Renewable Energy...and 'Green Energy / Clean Energy' Today!

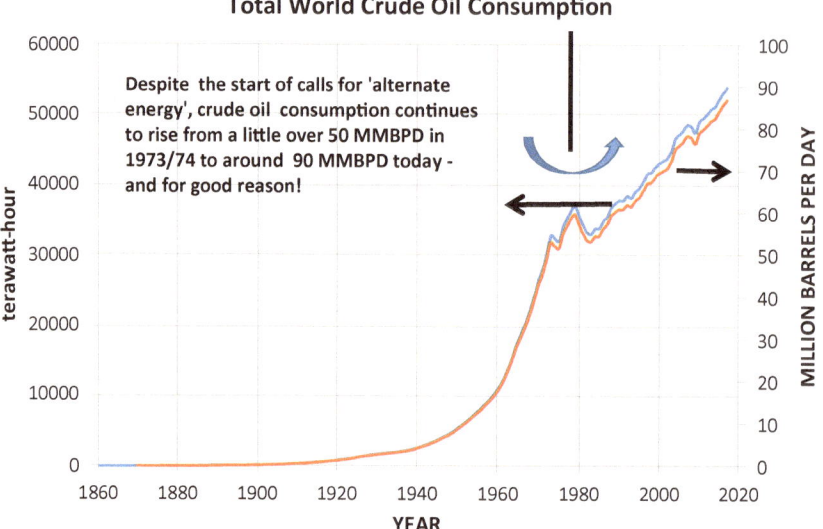

Fig. 1.1 Historical crude oil prices and production rates. *Data Source* Mainly EIA and US BLS; Smil (2017), energy transitions: global and national perspectives, ABC-CLIO, LLC

Clamor for alternate energy waxed and waned with fluctuating oil prices despite the fallacy pointed out above—however, a new business counter had opened up. Calls for 'alternate energy' (solar/wind/geothermal/others) became calls for 'renewable energy' (solar/wind/geothermal/others), with emissions from fossil fuels being held responsible for the global warming—and, now calls for 'green energy' (solar/wind/geothermal/others) to combat the 'climate change'. In fact, 'green energy' is not 'green' in today's context—it is also 'black' and the authors discuss this as appropriate in the following chapters.

'Climate Change' is real but whether it is all man-made or not is highly debatable. Earth has cycled through warming periods and ice ages through millions of years, hundreds of millions of years, and discussion of 'climate change' is beyond the scope of this book but we do touch upon it briefly towards the end of this book. This is not to say that humans should degrade the 'Environment'—modern life comes at a cost; there is no free lunch!

1.2 Charting the Course

The role of fossil fuels, and for good reason, is undeniable in the development of the West and modern day living. The debate over using 'alternative energy' ('renewable energy'/'green energy'), consciously or unconsciously, neglects the fact that it is not a replacement for fossil fuels. "While renewable resources gave us our start, renewables are also what we left behind."[1]

President Carter in his address from the Oval Office on July 15, 1979, said, among other things[2]:

".... calling for the creation of this nation's first solar bank, which will help us achieve the crucial goal of 20% of our energy coming from solar power by the year 2000." Solar energy contribution to the energy pie in the US today (2018) is less than 1.0%—as against the target stated by President Carter of 20% by 2000! Figure 1.2 presents the US energy consumption by type in 2018 and dependence on fossil fuels is clear. The US with about 5% of the world's population, uses about 20–25% of the world's energy to lead the lifestyle it does! One can only try and imagine what would happen if India and China with a third of the world's population were to provide a similar lifestyle to their people!

It is also important to understand that 'green energy' is not all that 'green'—and, requires 'black (fossil fuel) energy' to harness, transform into a usable form, and control its use to deliver the benefits of modern life. True, there have been various technological advances in different fields since the 1970s, and the world is not the same, but some things never change like the energy-basics.

---

[1] Darmstadter et al. (1983).

[2] https://www.washingtonpost.com/archive/politics/1979/07/16/text-of-president-carters-address-to-the-nation/70414c34-0ac4-4de1-b579-3af3937146d4/.

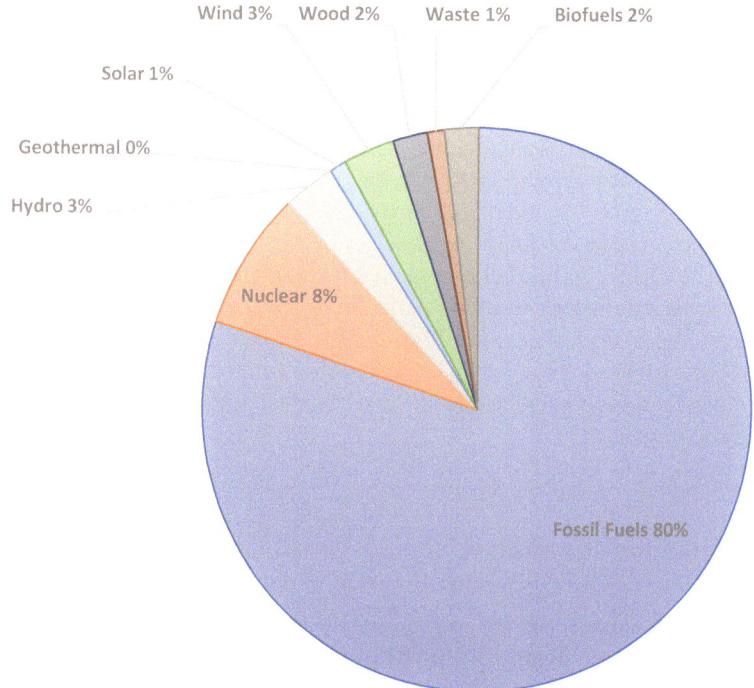

Fig. 1.2 US energy consumption by type in 2018. *Source* Monthly Energy Review April 2019

It is critical to sift hype from reality; and, it is not the intention of the authors to propagate the use of any one kind of energy over another but to present facts and discuss the entire energy space for the reader to arrive at informed conclusions. Energy fundamentals, and laws governing all matters energy which cannot be violated, are discussed first in 'English' in the next chapter.

Last, but not the least, to be provocative and leave the reader with some food for thought: Carbon is the fourth most abundant element in the Universe; is the essence of life on earth and in the form of carbon dioxide the only source of oxygen we breathe. Further, the human population on earth a hundred years ago was about 1.7 billion with a reported CO_2 concentration in the atmosphere of about 100 ppm; today (2020), the human population of earth is about 7 billion with a reported CO_2 concentration in the atmosphere of about 400 ppm!

References

Darmstadter, J., et al. (1983). *Energy today and tomorrow.* Englewood Cliffs, NJ: Prentice-Hall.
Smil, V. (2017). *Energy transitions: Global and national perspectives.*

Chapter 2
Energy Fundamentals in 'English'—*Understanding There Is No Free Lunch*

2.1 Energy—*Engine of Life*

Energy is the capacity to do work. From the mundane to the complex, all mechanical action is work. Raising an arm, lifting a glass, walking, riding a bicycle, running a car, flying an airplane, sailing a ship, operating a washing machine, running an air-conditioner, operating a kitchen mixer/grinder, manufacturing products of modern day conveniences, sending a space ship to the moon, launching satellites, and many, many more, are all mechanical actions requiring energy—different kinds of energy, but energy all the same! *It is not necessary for energy to do work—it may or may not—but it has the capacity to do work.*

Energy is required to live; and, creative usages and applications of the available natural resources over time have grown and facilitated a better life. Energy is the 'engine' of life! Later in the book, the authors track energy over time and its impact on human life. Also presented is the entire energy space with each form of energy discussed separately with facts and a perspective.

All life and energy on Earth are because of our Sun—take the Sun away (it should be around for billions of years yet) and life as we know it shall cease to exist! It is a different matter that human life can still be wiped out in an instant like what happened with the dinosaurs—another conversation for another day!

Thermal energy is at the heart of most mechanical actions, which are then gainfully employed, directly or indirectly, to provide comforts of modern life as mentioned above. Direct use of thermal energy (by way of fuels) is for transportation (surface—rail/road; air; and, sea) and certain manufacturing processes; while indirect use of thermal energy is by way of first producing electricity (by mechanical action) which is then distributed to billions of outlets in homes (to provide modern conveniences) and factories (for certain manufacturing processes). Energy is also used for heating, cooling and cooking. Ideally, one can do everything with thermal energy (available in the form of natural resources) that one can do with electricity, except that electricity in homes provides modern conveniences of life. Having said this, the authors are of the view that the biggest misuse of electricity—not a natural resource—is

© Springer Nature Singapore Pte Ltd. 2020
R. Sharma and V. Pareek, *The World of Energy*,
https://doi.org/10.1007/978-981-15-6724-7_2

for heating, cooling, lighting and cooking except that it provides a convenience! Figure 2.1 presents a schematic of the effect of energy on life.

Two other natural sources of energy are Potential Energy and Kinetic Energy. Potential Energy of an object (mass) is because of its position in space (height above the ground). Because of its position in space (height above the ground), an object has the potential to do work as it falls (gravitational acceleration); and, since capacity to do work is energy, the object has 'potential energy'. An object at rest on the ground has no 'potential energy'. Kinetic energy of an object is because of its motion. Suffice it to say here, that an object (mass) moving at a certain velocity has the potential to do work; hence, has 'kinetic energy', a term coined by Lord Kelvin.

Since 'Energy' drives 'Life', it is imperative to understand some basic laws (called the Laws of Thermodynamics—essentially, Laws of Nature) governing energy and its transformation as discussed below. These laws cannot be violated under any circumstances!

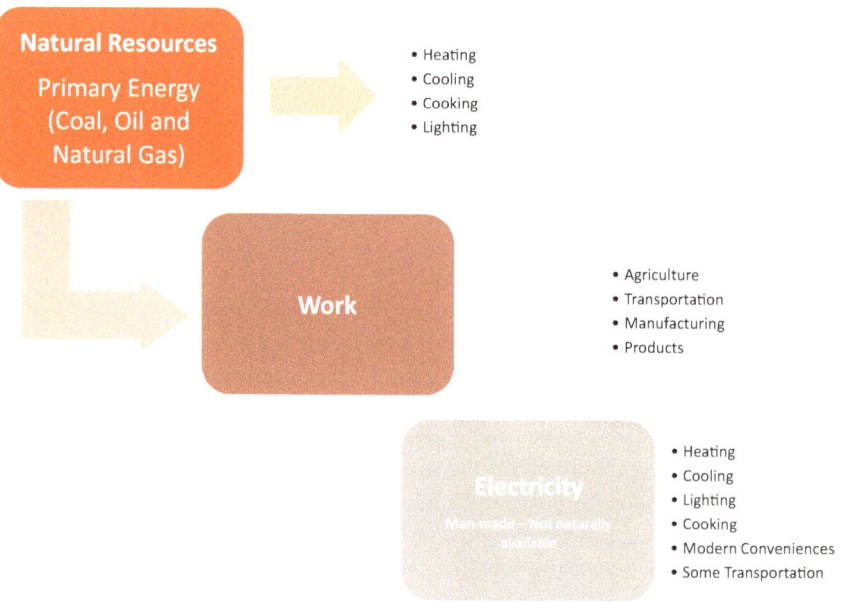

Note: Natural Resources of Coal, Oil, Gas are also used as feed material in the manufacture of industrial chemicals, petrochemicals, different products for various uses.

Fig. 2.1 Effect of energy on life

2.2 Energy Laws (of Thermodynamics) in 'English'—*There Is No Free Lunch*

Thermodynamics, in simple terms, may be explained as '*motive-power of heat*' (*thermo* is heat); *dynamics* is rate of mechanical action (work); and, rate of doing work is *power*.

All energy and alternate energy sources, without exception, **must** rigorously obey the laws of thermodynamics. All the laws of thermodynamics are empirical laws and are statements based on observations that cannot be proved but are laws of nature[1] (just as the law of gravitation). The three most important rules tell us that:

- Energy is conserved—you cannot get 'out' more than what you put 'in',
- All thermal energy cannot be converted into work—it relates to Carnot Efficiency (as shown in Fig. 2.2), and
- When left to themselves, systems tend to less and less useful forms in a never-ending quest to maximize their own disorder (also described as '*entropy*').

Ultimately, no matter how it is used in what way, virtually all energy ends up as useless low-grade heat.

Hence,[2] the above rules (laws) may be stated as

- you cannot win
- you cannot break-even either, and
- everything eventually goes to 'naught'!

One cannot do without energy—but there is no free lunch! Adapting a common saying, 'it's all about energy, honey'! Perhaps the answer lies in conservation rather than exploitation of energy resources. There may be a need to re-think the Western (US) model of development as one size fits all—particularly, the per capita usage of energy in the US to maintain that lifestyle seems to be unsustainable—what with the world aspiring for a similar lifestyle!

2.3 Quantifying Energy—*Getting a Measure*

Just as a unit of measure is required to quantify mass (kg; lb; …), length (m; ft; …), time (s; min; h), temperature (C; F; K; R), velocity (m/s; ft/s; mph; …), pressure (Pa; atm; psi; …), force (N; lb$_f$;), and many others, so it is for energy. The unit of measure of energy is joule (J; named in honor of James Prescott Joule; the other popular unit for energy is calorie (cal)), just as newton (N; in honor of Isaac Newton) for force, or pascal (Pa) for pressure or watt (W; J/s) for power. Power is the rate of doing work;

---

[1] Van Ness (1969).

[2] Adapted from Dugdale (1966).

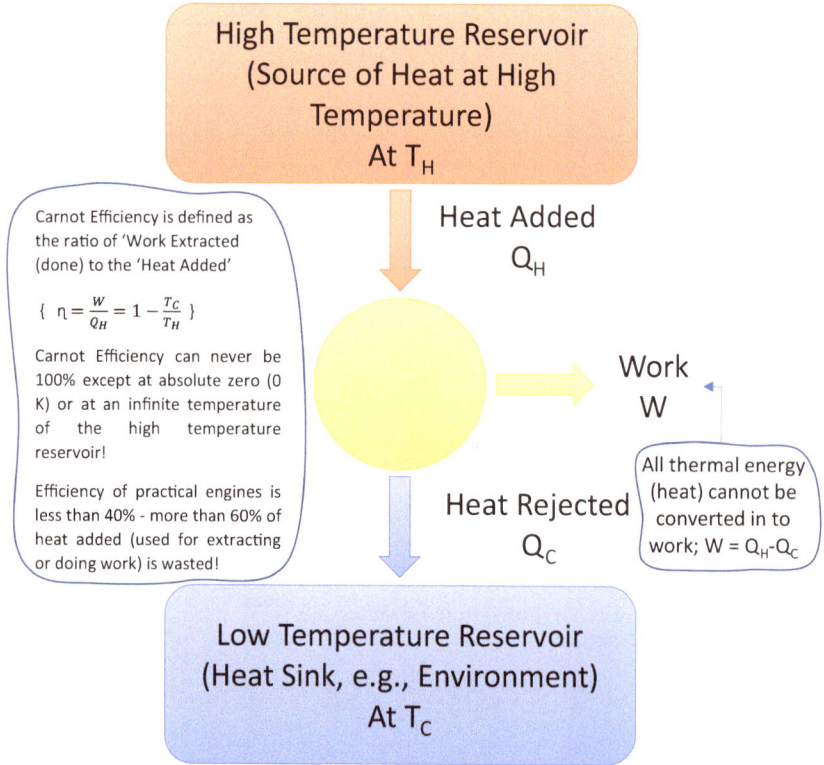

Fig. 2.2 Carnot engine

and, energy and work have the same units of measure. The following represents the commonly used units of measure and their equivalence.

Measures of Energy

1 kJ (kilojoule)	= 0.9478 Btu (British thermal units) = 238.8 cal (calories)
1 W h	= 3.6 kJ = 3.41 Btu
1 toe (ton of oil equivalent)	≈ 42 GJ (gigajoule) ≈ 40 million Btu ≈ 12 MW-hr (megawatt-hour) ≈ 7–8 boe (barrel of oil equivalent)

(continued)

(continued)

1 ton of coal equivalent	$\approx$ 29.3 GJ
	$\approx$ 28 million Btu
	$\approx$ 8 MW-hr
	$\approx$ ¾ boe
1 million toe	$\approx$ 12 TW-hr (terawatt-hour)
1 million ton of coal equivalent	$\approx$ 8 TW-hr

Measures of Weight

1 metric ton	= 1000 kg (kilograms)
	= 2204.62 lb (pounds)
1 short ton	= 1.103 short tons
	= 2000 lb (pounds)

Prefixes

1 k (kilo)	1000
1 M (mega)	1000 k
1 G (giga)	1000 M
1 T (tera)	1000 G

Practically then, to get a feel for these measures:

- 1 GW-hr of energy can light up about 100 homes for 3 years, assuming each home consumes about 8–10 kW-hr of energy each day; or, 1 TW-hr can light up about 100,000 homes for three years.
- 1 GW-hr of energy can cover about 1 million miles in a car averaging about 35 miles per gallon of petrol. If the average distance covered in a year is 25,000 miles, this energy is enough to drive the car for 40 years. 1 TW-hr can drive 1000 cars about one million miles.

2.4 Of Energy Sources, Sinks … and Carriers

We always need an 'older'(existing) form of energy to produce 'newer' energy—such as we mine (using some 'older' energy) coal ('newer' energy); we drill (using 'older' energy) oil and gas ('newer' energy); produce electricity ('new' energy) from coal or solar energy ('older' energy); and the like. When the ratio of the newer energy produced to the older energy consumed is greater than one, the newer energy is an *energy source* (such as coal, oil, and gas). Of course, the higher this ratio, better the energy source. *The original source of all energy (life!) on earth is the Sun!*

On the other hand, when the ratio of new energy produced to old energy used is less than one, new energy is an energy sink—as an example, production of electricity is an energy sink. In fact, one is paying for convenience in terms of energy (electricity) use. Different modes of electricity production not only have to compete in terms of cost but also energy used, scalability, safety hazards, material and resource availability and limitations; and, many other factors.

Energy carriers deliver energy from one point to another for use. Electricity is also an example of an energy carrier available for heating, lighting, and doing mechanical work at the point of use. Flywheels and batteries are some other examples of energy carriers. All energy carriers are indeed energy sinks.

2.5 Tail-Piece

With the above background, in the next chapter let us look at an over-view of the entire energy space before discussing each major energy type separately.

References

Dugdale, J. S. (1966). *Entropy and low temperature physics.* London: Hutchinson.
Van Ness, H. C. (1969). *Understanding thermodynamics.* New York: Dover Publications Inc.

Chapter 3
The World of Energy—*The Big Picture*

3.1 Order of Energy—*Man's Tryst with Fire*

It all began with fire hundreds of thousands of years ago; and, continues to be so today—control over energy! Man's discovery of fire, and his/her affair with it, led to improvements in his/her life over millennia by way of heating, cooking, and lighting—and this continues to be so even today; directly or indirectly. For fire, three things are essential—something that burns (fuel), oxygen, and a source of ignition—fuel and oxygen (air) were and are supplied by Nature; and, spark by Man. Intensity of fire is determined by the energy given off by the burning fuel (heat of combustion) and its rate of burning which, among other things, depends on fuel-type. Biomass (wood/plant material) was the first fuel (and still is a fuel in many parts of the world even today) but Man's quest for 'better' led to discoveries of coal and petroleum thousands of years ago. Coal was initially used for cooking and heating while petroleum (distilled) for lighting. Harnessing the energy of waters and winds was attempted a thousand years ago by building water wheels and windmills (wheel had already been invented a long time ago) and sails for ships—but there was no control over the source of energy and one had to accept what one got. Water and wind could not be ordered to flow and blow; neither their direction nor their velocity! But, humankind's thirst for a better life was unabated—fast forward to contemporary times in terms of earth time; about 300 years ago.

Although until early Eighteenth Century, the use of fire was limited to being a heating source (to keep warm or cook food), humans had started to realize the ability of heat to impart rotational motion since the beginning of First Century AD, when a Greek inventor named Hero designed the world's first aeolipile, or primitive steam turbine (Fig. 3.1). Hero's device consisted of a hollow sphere, mounted on a pair of tubes that transported steam to it from a heated pot underneath. The steam was then released through two bent tubes emanating from the sphere's equator which caused the sphere to revolve, thus demonstrating the potential of heat to generate rotational, and eventually translational motion.

© Springer Nature Singapore Pte Ltd. 2020
R. Sharma and V. Pareek, *The World of Energy*,
https://doi.org/10.1007/978-981-15-6724-7_3

Fig. 3.1 Aeolipile: demonstrating that heat could be converted to rotational motion

However, humans continued to rely on themselves, or domesticated animals (horses, camels, oxen, dogs, yak and others) to carryout 'work' until the early Seventeenth Century when driven by the need to pump out water from silver mines led a Spanish inventor named Jerónimo de Ayanz y Beaumont obtaining the first patent for a steam engine in 1606. However, it was only after James Watt invented the improved steam engine powered by coal, in the latter part of the Eighteenth Century, that the motive power of heat was well and truly established—and humans had direct control over energy. Mills, breweries, factories began to be powered by steam engines to do work at a speed and scale never seen before! Sprawling train networks with trains pulled by steam engines to transport passengers and material sprung all over Europe starting in early Nineteenth Century. Steam ships were built powering travel by sea. The first steam locomotive was imported in the US in the late 1820s and similar developments as in Europe were underway. Meanwhile, improvements in the steam engine design kept on progressing. Industrialization of the West was well on its way—all powered by improved quality and control of energy (coal, and steam engine).

Petroleum was discovered in the US in 1859 in any large measure and the first oil well drilled in Pennsylvania. A new and a better-quality energy source had been discovered; and, exploration and production of oil spread rapidly across the United States. Though the initial use of petroleum was as a replacement for whale oil in street lighting, invention of the internal combustion engine using petroleum (refined) as the energy source saw the rise of automobiles (and later airplanes/diesel locomotives/and bigger, better and faster ships). The concept of 'suburbs' was born in New York to push the use of automobiles as providing individual freedom. Exploration and production of petroleum in other parts of the world had also begun!

Other developments were also happening rapidly—invention of the light bulb and the earlier discovery of electricity saw the first electric street lighting in 1892 when GE installed electric generators in Manhattan to light up Pearl Street. Urban electrification was on and coal fired thermal power plants began being set up at a frenetic pace. Applications of electricity began springing up—as examples, air conditioning/refrigeration/pumping/many home and kitchen appliances—and living comforts were at hand.

Following the end of the Second World War, over the next twenty years or more, massive re-building of Europe and of the massive infrastructure in the US—transportation/communication/electric grid/etc.—happened on the back of cheap, abundant, and reliable energy; coal, oil and natural gas. It is interesting how GE pushed home appliances for better living by calling them 'electrical servants' in their famous advertisement "Live Better Electrically" in the 1950s starring the Reagans (Ronald and Nancy, who later went on to become the 40th President of the United States and the First Lady, in 1980). The developed world with modern comforts of life had arrived!

Fire began it all—and, fire is what keeps it going! Burning of better-quality fuel, and directed use and control of the energy released on burning (heat of combustion) is what drives the comforts of today. It is interesting to pause for a moment and think that it is 'carbon energy' that is driving life; all animal and plant life is carbon based; even the food we eat to provide energy for the body is carbon-based.

And, finally, our Sun is the cause of all causes—even the so-called fossil fuels are solar energy concentrated over millions of years and stored in a compact form available for any use, anytime, anywhere, at the 'touch of a button' ('glow of a spark').

3.2   Global Energy Consumption—*An Unquenchable Thirst!*

Directed use and control over better quality, cheap, abundant, reliable energy, available anytime, anywhere at the 'show' of a spark, has been the key to modern comforts of life. The years following the Second World War saw an explosion in energy consumption as shown in Figs. 3.2 and 3.3. Most of this massive increase in energy consumption happened in the West, particularly in the US where it almost doubled from 10,000 TW h in 1950 to almost 20,000 TW h in 1970 (Fig. 3.4). The total world energy consumption over the last 200 years has increased by thirty times from about 5000 TW h in 1800 to about 150,000 TW h in 2017 (Fig. 3.3). Figure 3.5 sketches

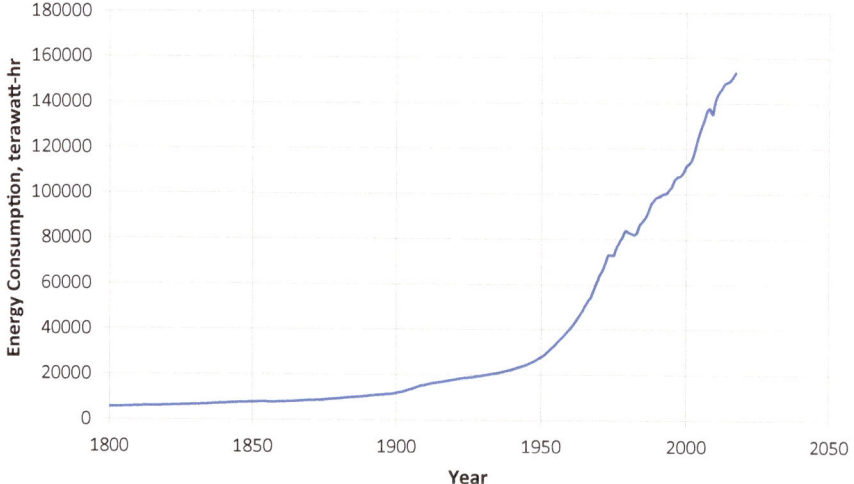

Fig. 3.2 Global energy consumption over the past two hundred years. *Data source* Smil (2017), energy transitions: global and national perspectives, ABC-CLIO, LLC

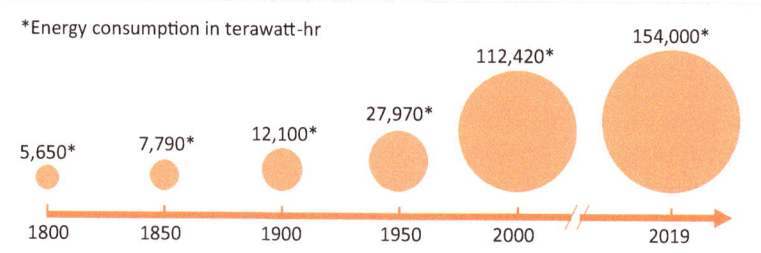

Fig. 3.3 Energy consumption and evolution of modern life over time. *Data source* Smil (2017), energy transitions: global and national perspectives, ABC-CLIO, LLC

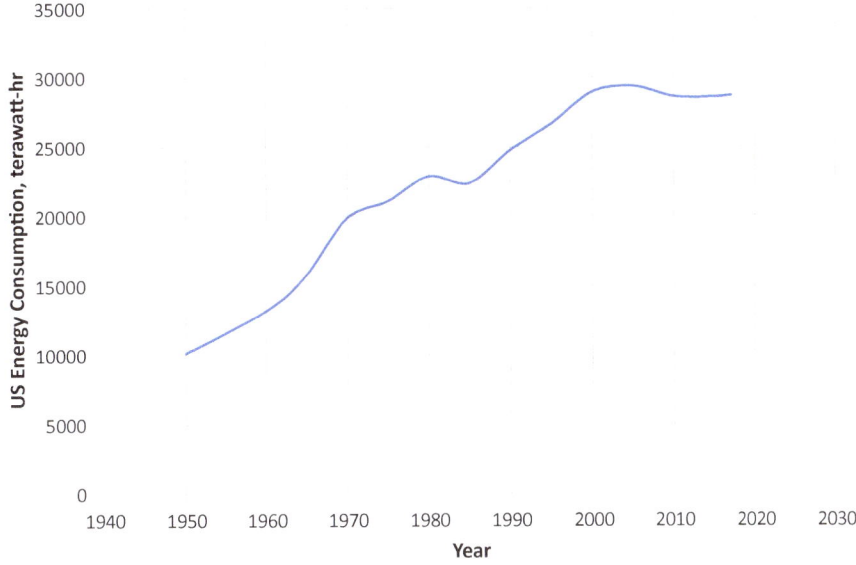

Fig. 3.4 Rapid increase in united states energy consumption since 1950—years of massive infrastructure development and modern-day living. *Data source* Smil (2017), energy transitions: global and national perspectives, ABC-CLIO, LLC

the evolution of energy use and the type of energy used from 1800 to 2017—in 50-year intervals—and the shift to coal, oil and gas is evident. Most of this increase has happened post World War II from about 27,000 TW h in 1950 to over 150,000 TW h today. The major difference between the so-called developed world and the developing world (in context of present day meaning of these words) is in the availability and use of energy. Even if energy was available, for example oil in the Middle East, it was mainly being used in the West.

Without energy, there can be no work; and, without work, there can be no productivity and economic growth! Winston Churchill, when he was the First Lord of Admiralty, is believed to have once said, after British engineers first struck oil in Iran (Persia) in the early 1900s, it all being about 'mastery of oil'. Taking the lead, today it all seems to be about 'mastery of resources'.

The whole world today looks up to the US as the model of development. Let us examine a few facts first: the per capita energy usage in the United States has stabilized to between 7000 and 8000 kg of Oil equivalent per person per year (around 85,000–100,000 kW h per person per year) over the past 35–40 years (peaking in 1978 to about 8500 kg Oil equivalent or ~105,000 kW h per person per year) to provide and sustain the lifestyle it leads (Fig. 3.6). In fact, the energy consumption in the US was already around 8000 kg Oil equivalent (~100,000 kW h per person per year) in 1972. In comparison, energy (oil) rich, tiny (in terms of population size) kingdom of Saudi Arabia consumed about 1000 kg Oil equivalent per person per year (~12,500 kW h per person per year) in 1972; whereas the 'giants' (in terms of

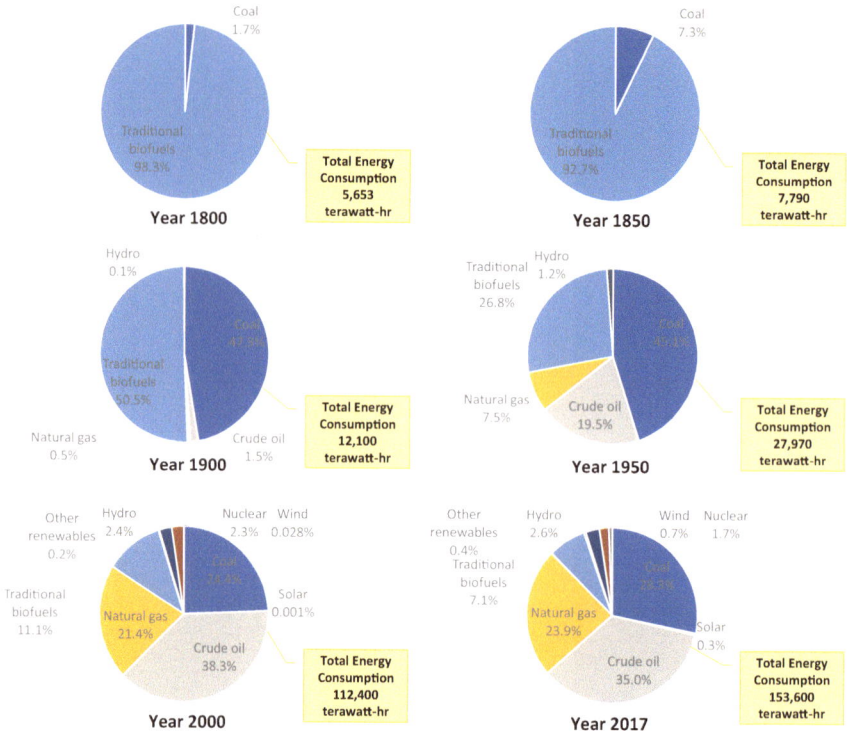

Fig. 3.5 Evolution of energy-type of world energy consumption. *Data source* Smil (2017), Eenergy transitions: global and national perspectives, ABC-CLIO, LLC; and BP statistical review of world energy 2018, 67th edition

population size) China and India consumed only about 500 kg Oil equivalent per person per year (~6000 kW h per person per year) and 270 kg Oil equivalent per person per year (~3500 kW h per person per year), respectively (Fig. 3.6). In other words, energy consumption in the United States in 1972 on a per capita basis was about 8 times that of Saudi Arabia; about 17 times that of China; and, about 30 times that of India. What is so special about 1972? That was the year of the 'ping-pong' diplomacy when President Richard Nixon, 37th President of the United States, opened the doors of the 'bamboo curtain' and brought China into the world-fold!

Again, the growth (development in the Western mold) of China (though it still lags far behind the US in terms of the entire country except for some major urban centers) propelling it to be a 14-trillion dollar economy, second only to the US (20 trillion dollar economy), is directly attributable to 'energy' with its consumption quadrupling over the last 35 years since the early 1980s to about 28,000 kW h per capita in 2014/15. This energy consumption of China is still far short of the US per capita energy consumption of 100,000 kW h per person in 2014 (Figs. 3.6 and 3.7).

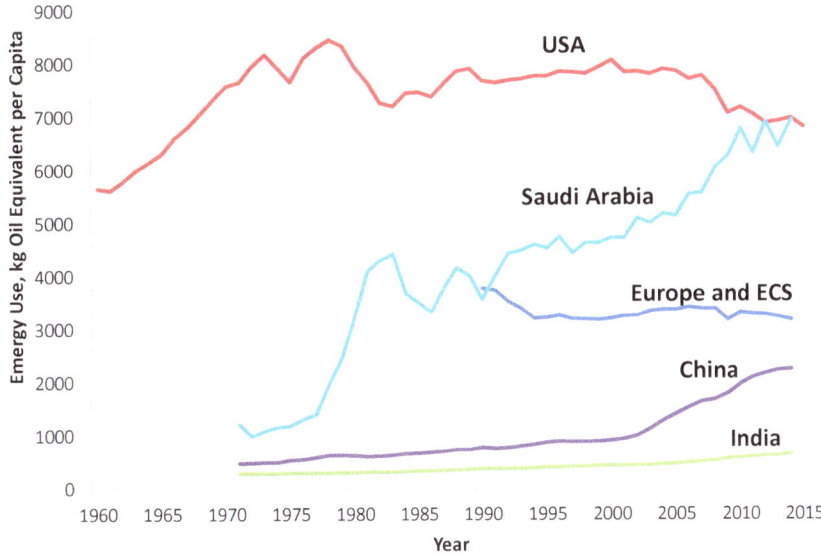

Fig. 3.6 A representative comparison of per capita energy use for various countries/regions with that of the united states. *Data source* World development indicators (last updated April 24 2019), the world bank

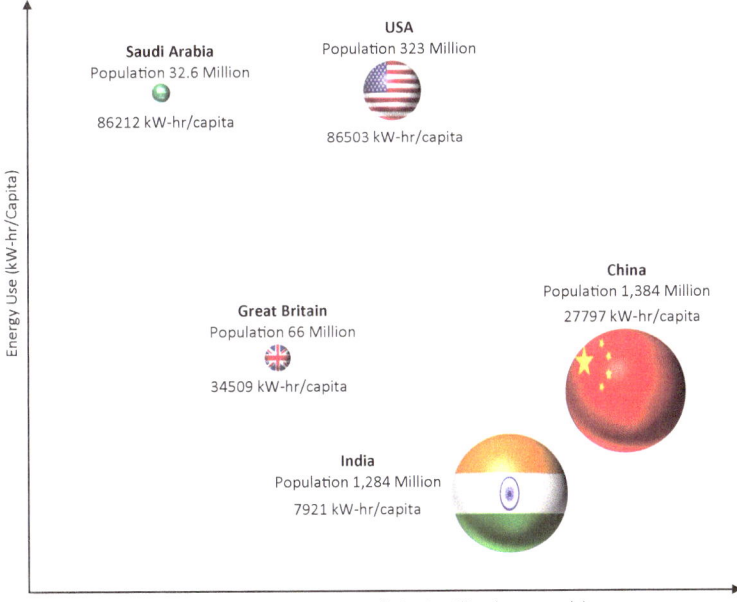

Fig. 3.7 A representative snapshot of energy use per capita for some countries. *Data Source* World development indicators (last updated April 24 2019), the world bank

A comparison of energy consumption for five countries, as an example—USA, China, India, Great Britain, and Saudi Arabia—presented in Figs. 3.8 and 3.9 is

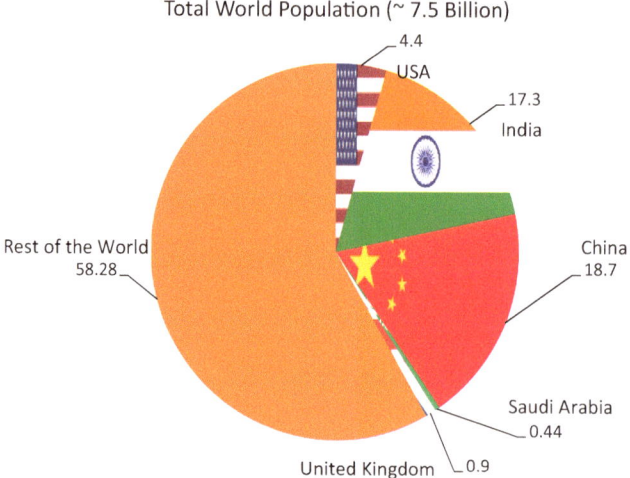

Fig. 3.8 Contribution in population of some countries as the percentage of total population. United states and china account for about 25% of the world's population. *Data source* World development indicators (last updated april 24 2019), the world bank

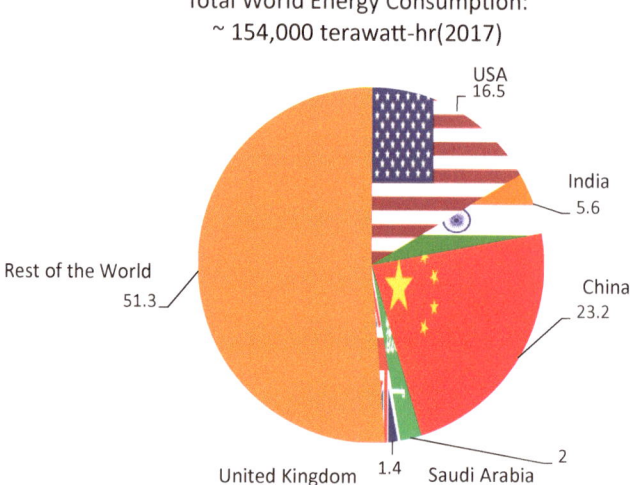

Fig. 3.9 Energy uses of some countries as a percentage of total energy. United states and china account for about 40% of world's energy use. *Data source* Smil (2017), energy transitions: global and national perspectives, ABC-CLIO, LLC

quite stark and compelling! USA, with less than 5% of the world's population, consumes about 16% of world's energy; with China at about 19% of the world's population consuming about 23% of the world's energy—both together, at about a quarter of the world's population, consume about 40% of the world's energy. India, with about 17% of the world's population, consumes less than 6% of the world's energy. Saudi Arabia, with about 0.5% of the world's population, consumes about 2% of the world's energy—it is interesting that not only does Saudi Arabia's per capita energy consumption matches that of the US, but also the ratio of per cent of world's energy consumption to the per cent of world's population is approximately the same. Great Britain, with about 0.9% of the world's population consumes about 1.5% of the world's energy—but it is very interesting to note that were the energy consumption of Great Britain to increase, on a per capita basis, to the levels of Saudi Arabia and USA, the ratio of their energy consumption to their population on a percentage basis would approach that of Saudi Arabia and the United States.

China surpassed the United States as the single largest consumer of energy in the world in 2009 with the present consumptions (2017) presented in Fig. 3.9, even though it has not yet reached the level of the United States in terms of energy consumption per capita (Figs. 3.6 and 3.7). China's rapid development and economic growth over the past 35-40 years, has happened on the back of cheap, abundant, reliable energy, available anytime, anywhere—coal, oil and gas, which meet about 87% of its energy needs (Fig. 3.10). India, another giant in terms of population size, is beginning to stir; and, even though its energy consumption is less than 6% of that

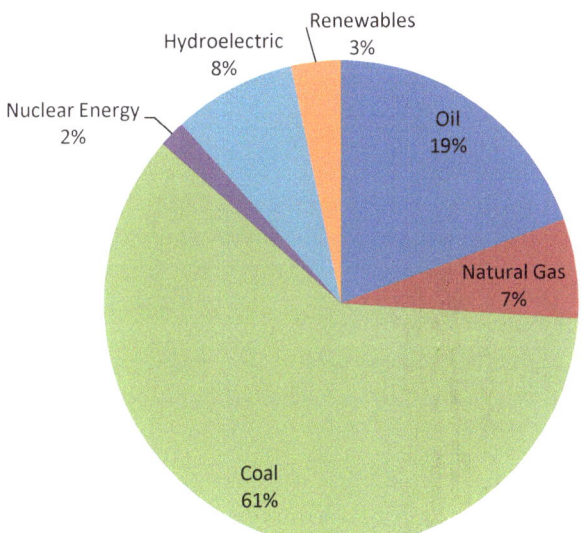

Total Energy Consumption: 3132 million tons oil equivalent

Fig. 3.10 China: Distribution of energy-type. *Data source* Smil (2017), energy transitions: global and national perspectives, ABC-CLIO, LLC

Fig. 3.11 India: Distribution
of energy-type. *Data source*
Smil (2017), energy
transitions: global and
national perspectives,
ABC-CLIO, LLC

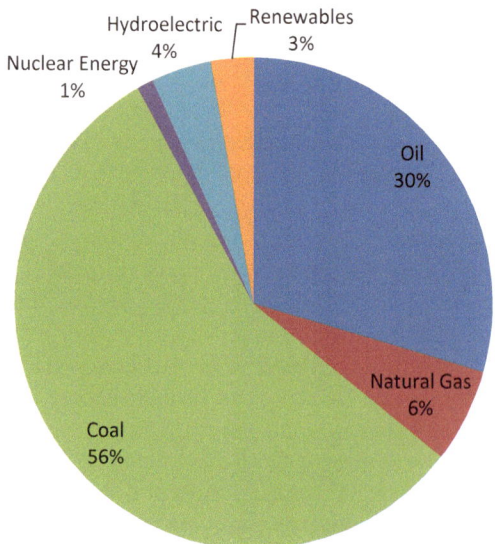

Total Energy Consumption: 754 million tonnes equivalent

of the world, 92% of its current energy needs are met by coal, oil, and natural gas (Fig. 3.11). Having said this, each type of energy shall be discussed separately in the following chapters, with the reader left to form his own opinions and arrive at his own conclusions.

As things stand in the world today, the thirst for energy shall continue to be unabated and unquenched for many decades to come—Fig. 3.12 shows a glimpse of this by presenting some projections till the year 2035. Figure 3.13 presents the world energy demand by fuel-type in 2035—coal, oil, and gas will continue to dominate with little variation from the present.

All this raises a few questions: Does the Western (US) model of development fits all? Is it sustainable for the world as a whole? Is it sustainable in the long run for the US itself? There are no easy answers—and the authors certainly don't have them!

Perhaps, just perhaps, the answers lie in examining the ancient civilizations of India, Egypt, and China, which were certainly rich and prosperous during their time, learning from them and combining ancient knowledge with modern science and technology, to find a path forward for a sustainable future of peace and prosperity for the world—albeit driven by energy.

3.3   Energy Space—*A Bird's Eye-View*

Figure 3.14 presents a quick overview of the entire energy space segmented into two major portions—Primary Energy and Alternate Energy, as shown.

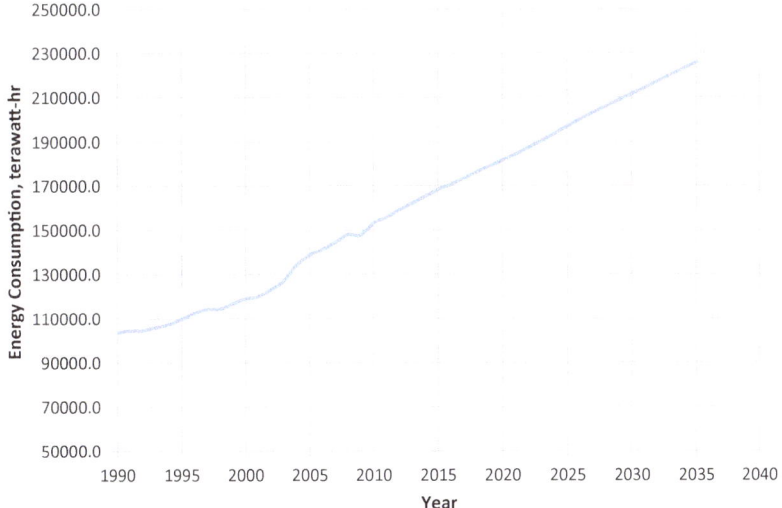

Fig. 3.12 World energy consumption and projections to 2035. *Data source* Adapted from us energy information administration international energy outlook 2011 doe/eia-0484(2011) september 2011; and bp energy outlook 2035, january 2014

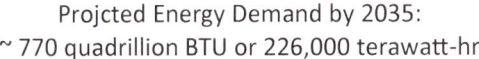

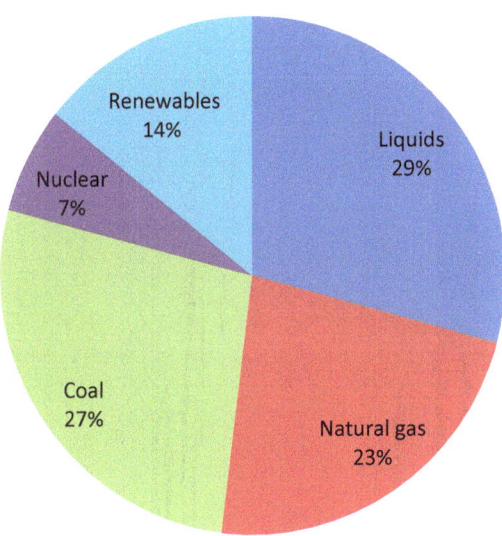

Fig. 3.13 Distribution of world energy demand by type–2035. *Data source* Adapted from bp energy outlook 2035, january 2014

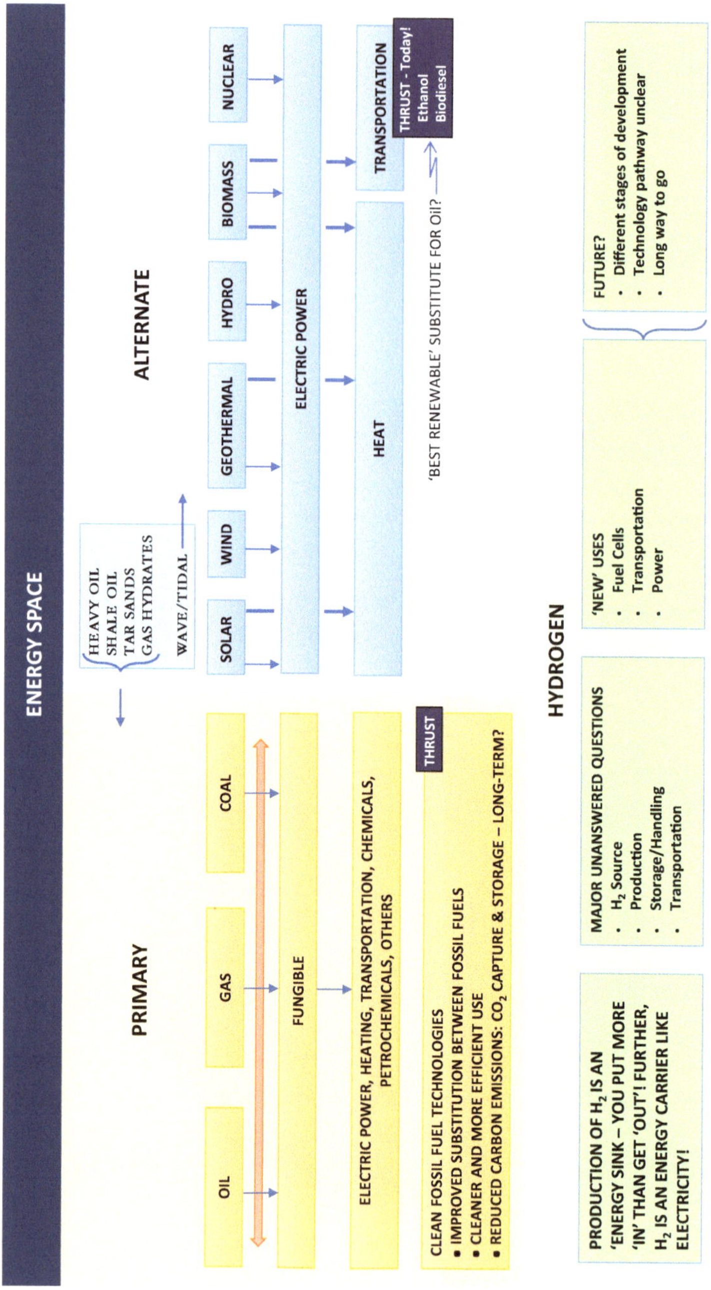

Fig. 3.14 An overview of the energy space

Primary Energy consists of coal, oil and gas. One of the major advantages of coal, oil and gas, in addition to the few discussed earlier, is that they are directly fungible—meaning that what one can do with coal, can be done with oil and gas; and, either which way! As an example, Germany was meeting a major portion of their requirements of transportation fuels, particularly during the War years (WW II), by converting coal into synthetic liquid fuels. South Africa, during the apartheid years for decades until the mid-1990s, when it did not have access to world's crude oil and gas, produced synthetic liquid fuels from coal using German technology—cars had to run, airplanes had to fly, work had to be done; and, it still produces liquid fuels from coal to meet part of its requirements of transportation fuels. Of course, coal has been, and still is, the mainstay for electricity production. Gas is mainly used for heating and electricity production while oil for transportation. In addition to serving as energy resources, coal, oil and gas are also a source of many industrial chemicals and their derivatives to make products useful for modern day life.

Alternate Energy, except geothermal and biomass, on the other hand, is not fungible and provides electricity as the single major product. Geothermal energy is mainly used as a low-grade heat source (with some exceptions for electricity production), while biomass (different forms of biomass) is used mainly as a source of low-grade heat and to produce biofuels (ethanol; biodiesel; these will be further discussed in a later chapter). Further, 'alternate energy' is not a source for any chemicals or by-products used in today's life. Biomass is the only truly renewable source of energy—but that is also carbon-based. Electricity is clean at the point of use but that may not really be so because of the requirements of its generation—even from solar and wind. As an example, generating electricity from solar energy requires photo-voltaic panels. If these solar photo-voltaic panels were freely available lying around, then perhaps solar electricity could be seen to be completely clean. But that is not so—photo-voltaic panels must be manufactured using 'old' ('black') energy and certain steps in the manufacturing process are a source of water pollution. Yes, solar and wind are clean if we go back to the old ways of life—nothing comes for free!

Energy by type, under both categories of 'Primary' and 'Alternate', are discussed in detail in the following chapters.

Reference

Smil, V. (2017). *Energy transitions: Global and national perspectives.*

Chapter 4
Primary Energy—*The Big Three*

4.1 A Quick Recap

Coal, Oil and Gas are the big three of 'energy'; and, for good reason. Together, they meet about eighty-seven percent of the requirements of world's energy (Fig. 4.1). Not only are they cheap, abundant and reliable sources of energy, when you carry them, you are carrying energy readily available for use anytime, anywhere, and for whatever reason, 'instantly'. Coal, oil and gas, as mentioned in an earlier chapter, are directly substitutable for each other implying any one of them can be substituted for any of the other for the same use—cooking, heating, transportation, electricity production, and chemicals production, as examples. Further, all three are natural resources and not man-made—sources of energy directly under Man's control. In effect they may be considered as solar energy stored in a carbon form (hydro-carbon) compacted over millions of years available in a readily usable form—after all carbon had to come from somewhere, it is a natural element and not man-made!

4.2 Coal—*The Workhorse*

Coal is the original 'Black Gold', formed from dead plant material over millions and millions of years—in fact it is nothing but solar energy stored in a compact and readily available/usable form. Discovered thousands of years ago, and first mainly used for cooking and heating, combined with the invention of the steam engine in the latter half of the Eighteenth Century as its source of energy, Coal became the workhorse for Man. Steam engines were deployed in factories first all across the United Kingdom, with rest of Europe playing catch-up, allowing Man to do 'work' on a speed and scale never seen before! Steam engines were deployed not only in factories all across Europe but also became a major source of power for transportation—train/railroad networks sprang up and European Navies, powered by coal-fired steam engines, became stronger and faster extending their colonial

© Springer Nature Singapore Pte Ltd. 2020
R. Sharma and V. Pareek, *The World of Energy*,
https://doi.org/10.1007/978-981-15-6724-7_4

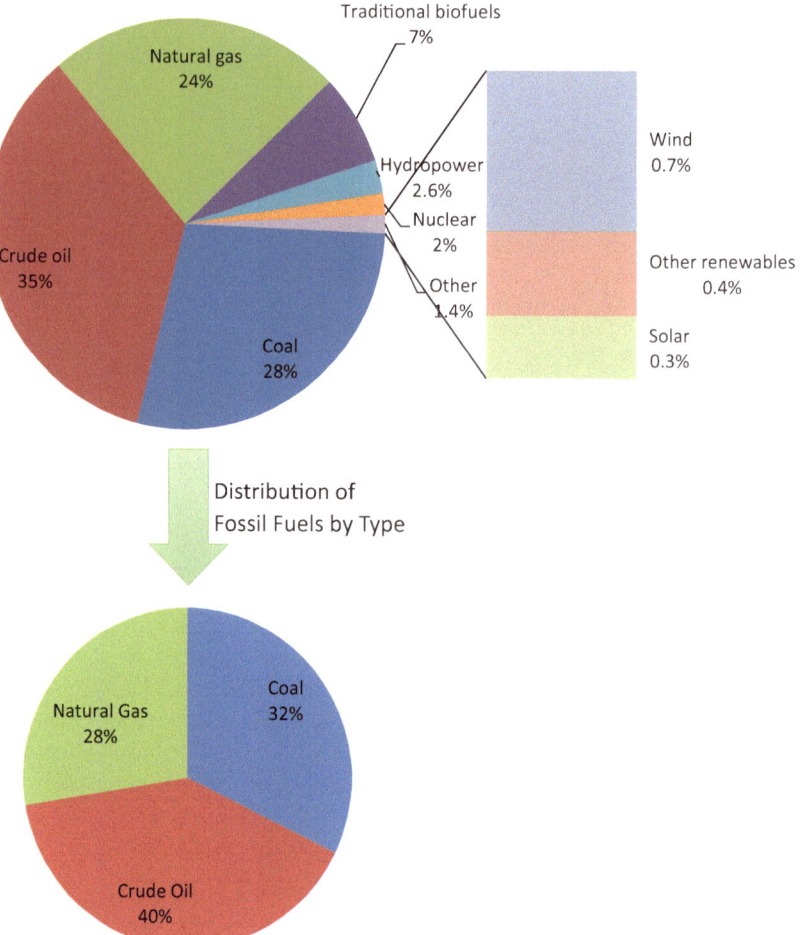

Fig. 4.1 Global energy consumption and distribution by type–2017. *Data Source* Smil (2017), energy transitions: global and national perspectives, ABC-CLIO, LL; and BP Statistical Review 2018

influence around the world. Steam engine crossed the Atlantic to the United States in the first half of the Nineteenth Century—and Industrialisation of the West was happening on the back of coal—a cheap, abundant, reliable, natural source of energy! Not only on the back of coal, which was of course the workhorse, but also on raw materials from the 'colonies'; which served as 'feed' for the factories in Europe to produce finished products to satisfy ever growing demands of the Western lifestyle.[1]

Electricity had been discovered in late Eighteenth Century. Invention of the electric light bulb in late Nineteenth Century, combined with the 'steam engine', led to sprawling thermal power plants being set up all across Europe and the United States for electricity production; and, electrification of the West was well and truly underway in the first half of the Twentieth Century—all powered by coal. The first coal-fired thermal power plant was set up in London in 1882. Of course, some thermal power plants, few and far between, began springing up in the 'Colonies' across the world— but nothing like what was happening in the West. Electricity, in essence is an energy carrier (as explained in Chap. 2) and several applications began being invented for ease and comfort of Man's life at home and work,—many of which we take for granted today. Discovery of Oil and invention of the internal combustion engine added a new dimension to ease and comfort of transportation—more on this in the next section.

What we call development today, and consumerism, happened mainly in the West during the first seventy-odd years of the Twentieth Century—and all this happened on the back of cheap, abundant, reliable, natural sources of energy readily available for use anytime, anywhere, for any purpose. All this while, most of the rest of the World struggled to catch up—and still does to emulate the Western model of development.

To further emphasise the role of coal in 'industrialisation', it is educational to look at Figs. 4.2 and 4.3. As an example, the similarities between the rates of coal production (and consumption) in the United Kingdom and the United States during the latter half of the Nineteenth Century with that of China during the last thirty years are stark, to say the least (Fig. 4.2)! Of course, the amount of coal production (and consumption) in China, because of its sheer size and population, far exceeds that of either the UK or USA. India is also now on the 'march'. It is also interesting to see that coal consumption on a per capita basis in the United States began declining as it increased in China in the late 1990s before finally falling below that of China around 2012 (Fig. 4.3)—this can be linked mainly to an increase in 'manufacturing' in China and a reduction in the United States.

---

[1] "In 1800 Europeans occupied or controlled about 34% of the land surface of the world; by 1914 this had risen to 84%."

"Britain led the nineteenth-century takeovers and ended the century with the largest non-contiguous empire the world has ever known. ("The sun never sets on the British Empire," as the British liked to say.) Britain exerted great influence in China and the Ottoman Empire without taking over direct rule, while in India, Southeast Asia, and 60% of Africa, it assumed all governmental functions." The Industrial Revolution, https://www.khanacademy.org/partner-content/big-history-project/acceleration/bhp-acceleration/a/the-industrial-revolution.

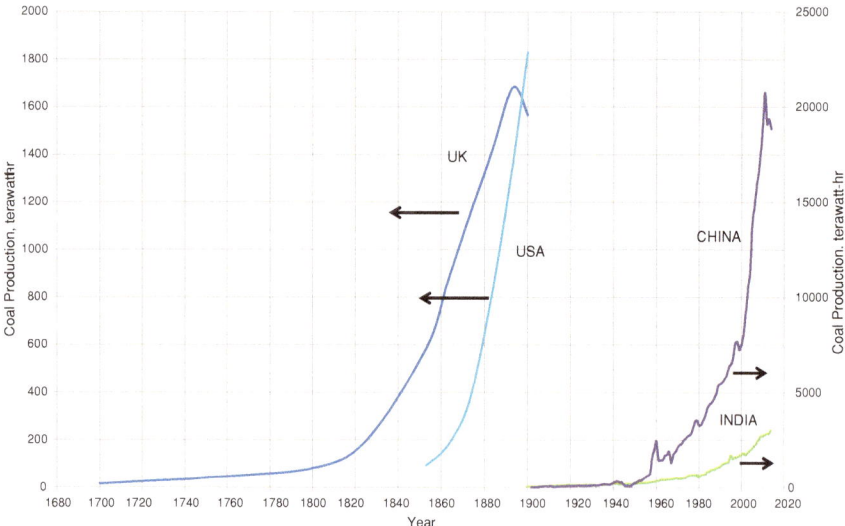

Fig. 4.2 Importance of coal in industrialization. *Data Source* The SHIFT project data portal; and, UK department of energy and climate change

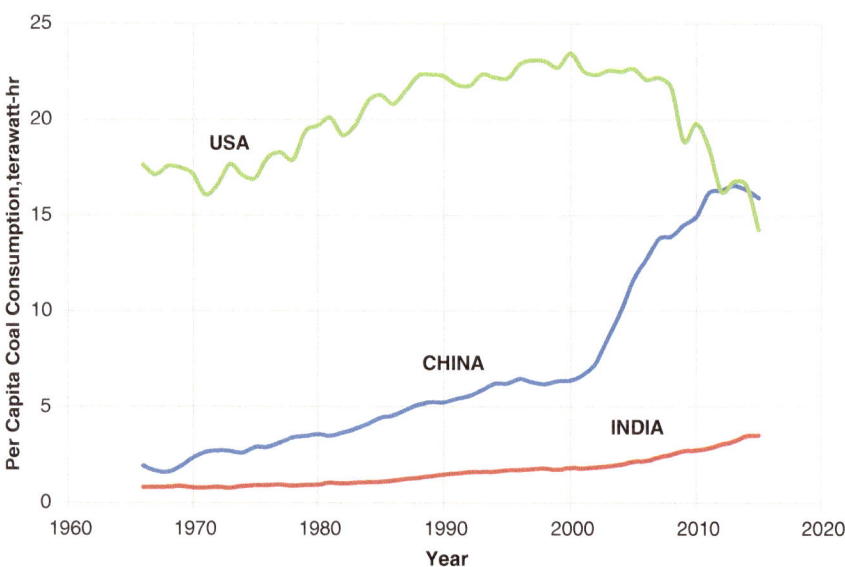

Fig. 4.3 Comparison of per capita coal consumption for a few major countries. *Data Source* Coal—BP Statistical Review of world energy 2017/UN population division (2017 revision)

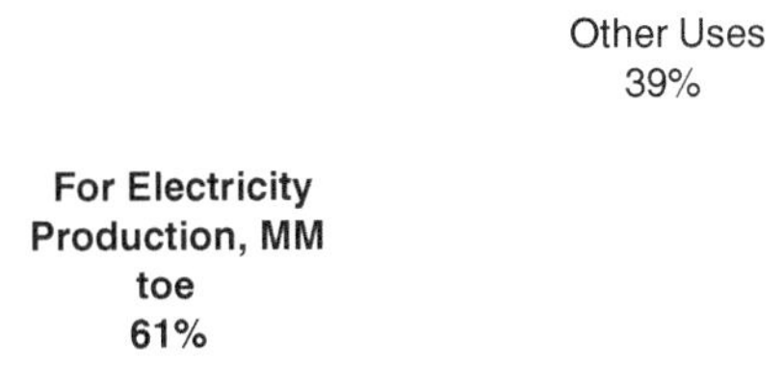

Fig. 4.4 Total world coal consumption, million tons of oil equivalent—2018. *Data Source* Adapted from BP Statistical Review 2019

Other Uses
39%

For Electricity
Production, MM
toe
61%

As things stand in the world today, there really is no substitute to coal for rapid industrial growth, particularly in terms of electricity production and manufacturing of cement and steel—all major requirements for development of modern infrastructure (Figs. 4.4, 4.5 and 4.6).

4.3   Oil—*The Mr. Big*

Oil is the 'Mr. Big' of the three major sources of primary energy and the new 'Black Gold'—making the world go around! It is 'Mr. Big' for good reasons—in addition to the energy punch that it packs compared to coal on an equivalent weight basis, its ease of handling, storage and transportation, have made it wrest the crown of 'Black Gold' from coal. Liquids, in general, are much easier to handle, store and transport compared to solids or gases. Although oil, like coal, was also discovered and first used thousands of years ago, it was only after its discovery in any appreciable quantities when the first oil well was drilled in Pennsylvania in 1859, that oil began playing an important and an ever-widening role in Man's life starting with lighting of street lamps (using distilled petroleum). Combined with the invention of the internal combustion engine during the latter part of the Nineteenth century, oil revolutionized the manner of transportation—land, sea and then air! Karl Benz produced the first automobile in the 1880s; Winston Churchill as First Lord of Admiralty (later to become the War-time Prime Minister of Britain), took an almost unilateral decision to switch from coal to oil as the energy source for the British Navy in the early 1900s after British engineers discovered oil in Iran in 1906, World War I saw the use of the internal combustion engine powered by oil; coal-fired steam engines for

Fig. 4.5 Total world
electricity production by fuel
type—2018: ~ 26615
terawatt-hour. *Data Source*
BP Statistical Review 2019

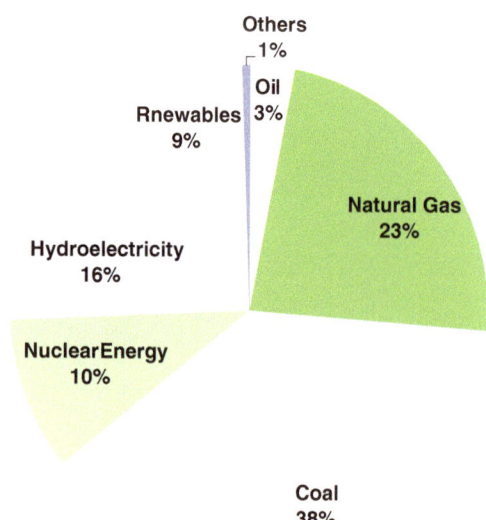

trains/rail-roads began being replaced—and, the world was starting to get smaller and closer!

Figure 4.7 presents the consumption of oil since 1860 after oil was struck in Pennsylvania in 1859. Oil consumption increased almost exponentially in the West (particularly in the US), in roughly 35 years, from just over 4 million barrels per day in 1940 to over 52 million barrels per day in the early 1970s—a period of rapid development in the West following the War years. And, then the Arab oil embargo happened following the Arab-Israeli Yom Kippur War in the early 1970s—but by then most of the infrastructure development in the West had already taken place with so-called modern lifestyle well on its way! In contrast, till 1940, even eight decades after oil production began on a meaningful scale, oil consumption reached just a little over four million barrels per day. With about 5–6% of the world's population, to fuel its lifestyle, United States has consistently consumed 15–20 million barrels of oil per day even over the past 45 years since 1975—a major share of the world's production ranging between 20–25% of the total!. In contrast, even though China today is the second largest economy of the world, at about 60–70% that of the US economy, its oil consumption is still about 13% of that of the world. With India also on the growth path in the Western mode and still using less than 5% of the world's consumption of oil, its oil consumption is bound to grow. On a per capita basis, United States has consistently consumed around 20–25 barrels of oil per person per year since the mid-1960s; while, China and India even today (2019) consume around 4 and little less than 2 barrels of oil per person per year, respectively (Fig. 4.8). One can only imagine the impact on the 'oil world' if China and India, giants in terms of their population size, were to replicate the Western lifestyle.

On a broader note, today more than half the oil is used to power the World's transportation needs, with about a third used for chemicals production, about six to

(Total World Electricity Consumption ~ 26615 terawatt-hour)

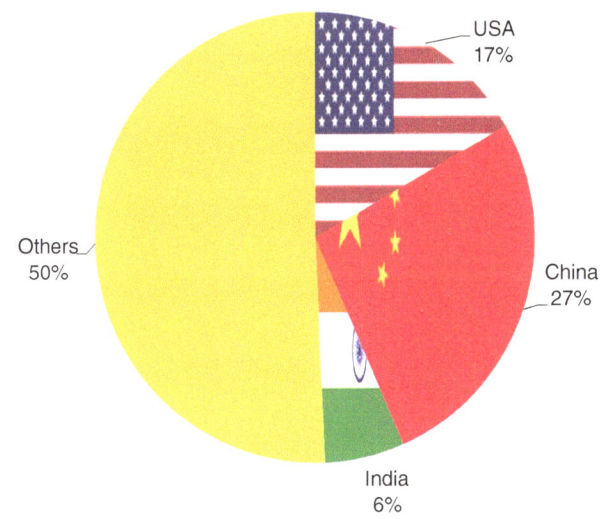

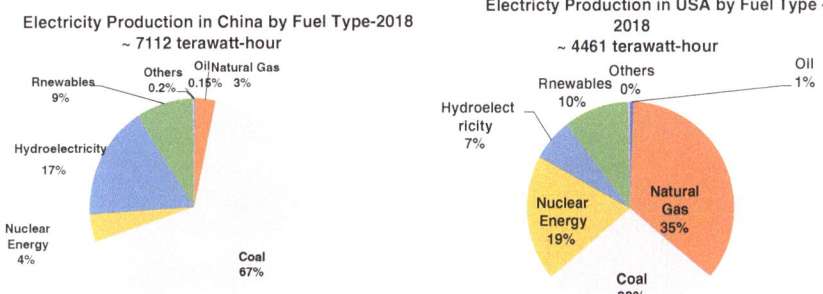

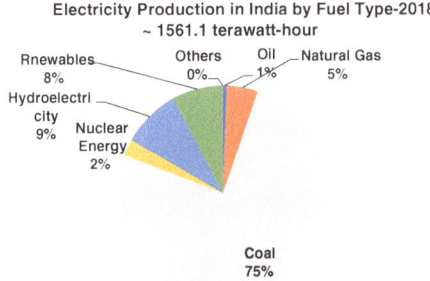

Fig. 4.6 Electricty consumption by three major countries—2018. *Data Source* BP Statistical Review 2019

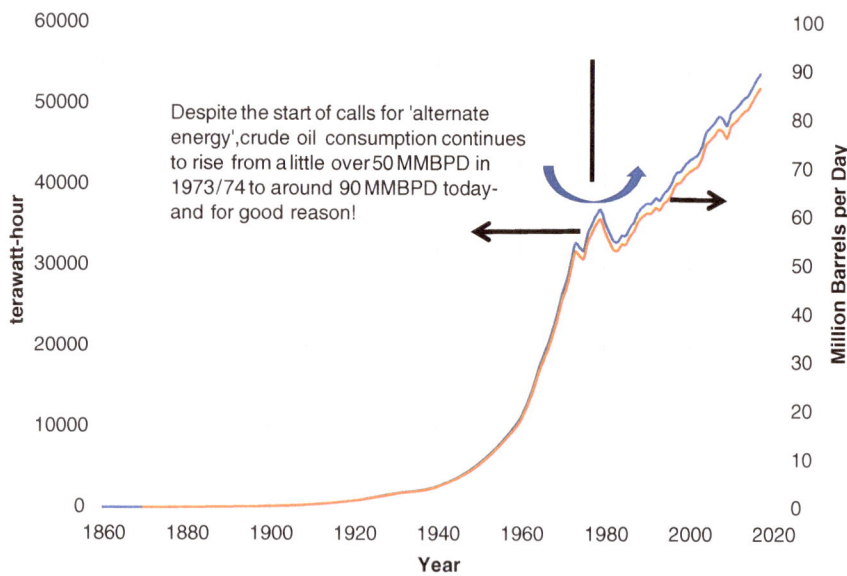

Fig. 4.7 Total world crude oil consumption (*Data Source* Smil (2017), energy transitions: global and national perspectives, ABC-CLIO, LL); and BP Statistical Review 2018)

seven percent used for electricity production (mainly in the Middle East), and the rest for heating purposes (Fig. 4.9). Oil drives the modern way of life! From providing gasoline, kerosene (aviation turbine fuel) and diesel (the three major products), oil is a key ingredient in thousands of products that affect our everyday lives and some examples are listed below (*source: American Petroleum Institute*):

Antiseptics	Aspirin
Cameras	Balloons
Bandages	Toothbrush/toothpaste
Inks/dyes	Tubes/tires
Clothing/candles	Fertilizers
Credit cards	Dinnerware
Eyeglasses	Paints
Diapers	Crayons
Furniture	Garbage bags
Tents	Heart valves
Computers	Lipstick
Insecticides/pesticides	Surgical/medical equipment

The list is almost endless!

4.4 Natural Gas—*Also a Gift of Nature*

Natural Gas, as the name implies, is also a natural resource available for ready and 'instant' use. Much like coal and oil, it is also a gift of Nature and not Man-made—the third constituent of the fossil fuel triad. The three natural states of matter are solid, liquid and gas—and fossil fuels are available in all three states; coal as a solid, oil as a liquid, and natural gas. Ever since its discovery in appreciable quantities, consumption of natural gas has increased many fold from about one-tenth of a million barrels of oil equivalent in 1900 to over 60 million barrels of oil equivalent in 2019 (Fig. 4.10). Till 1975, the United States accounted for over 50% (till 1965 it was over 70%) of the world's consumption of Natural Gas—again during the 30-year period of rapid development and economic growth following the War years (Fig. 4.11). Even today, the United States continues to dominate the consumption of Natural Gas having stabilized at about 20–25% of the world's consumption (Figs. 4.11 and 4.12)—in absolute terms, USA's consumption has almost doubled from about 4600 TW-hr in 1965 to about 8500 TW-hr in 2018. As mentioned earlier, there is no substitute for cheap, reliable and abundant energy for development and economic growth of any country in the Western mode—and China and India, with their burgeoning populations are trying to catch up. The question remains: Is the Western model of development one size fits all?

Natural gas has also found many uses to support 'modern life' (Fig. 4.13), including as a substitute for coal for producing electricity where it is readily available. Gases, in general, are relatively difficult to store and transport in comparison

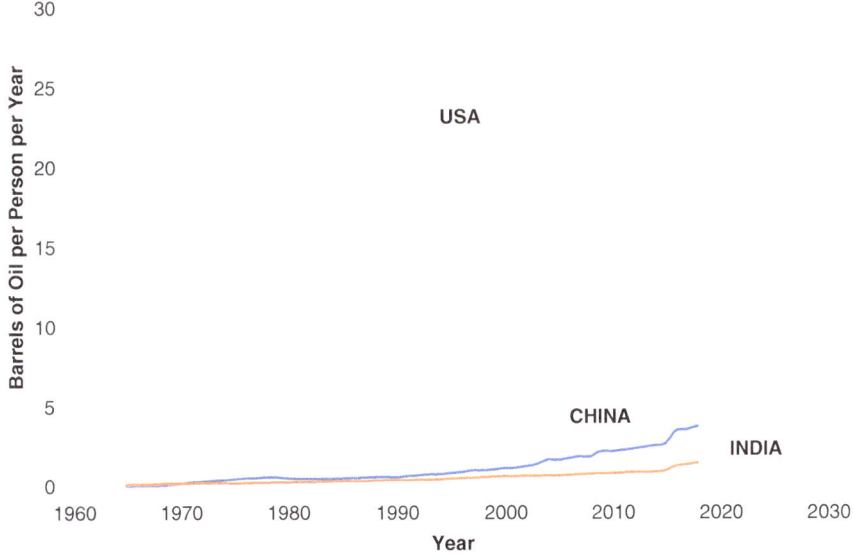

Fig. 4.8 Comparison of oil consumption per capita—China, India and USA. *Data Source* World development indicators (last updated April 24 2019), the world bank

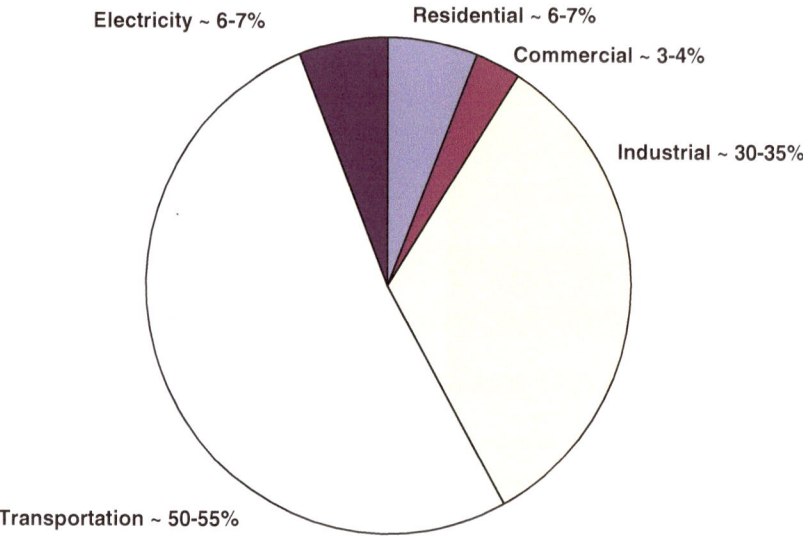

Fig. 4.9 Oil usage by sector—current to 2030. Data gleaned from several sources—EIA, IEA, WEO. Also, the percentage contribution of fossil fuels towards the total energy consumption, the distribution amongst them, and their sector-wise usage is not expected to change much over the next 25 years or so from the current ranges

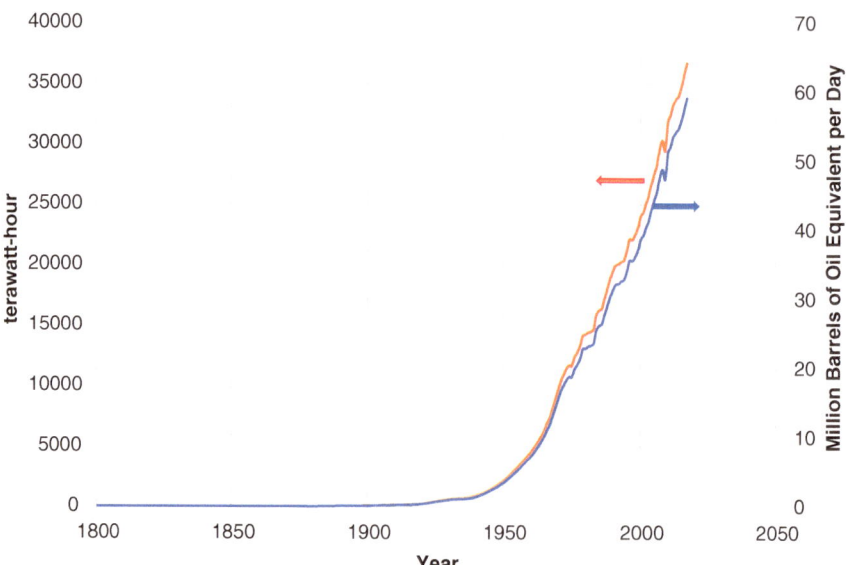

Fig. 4.10 Natural gas consumption over the past 100 years. *Data Source* Smil (2017), energy transitions: global and national perspectives, ABC-CLIO, LL); and BP Statistical Review 2018

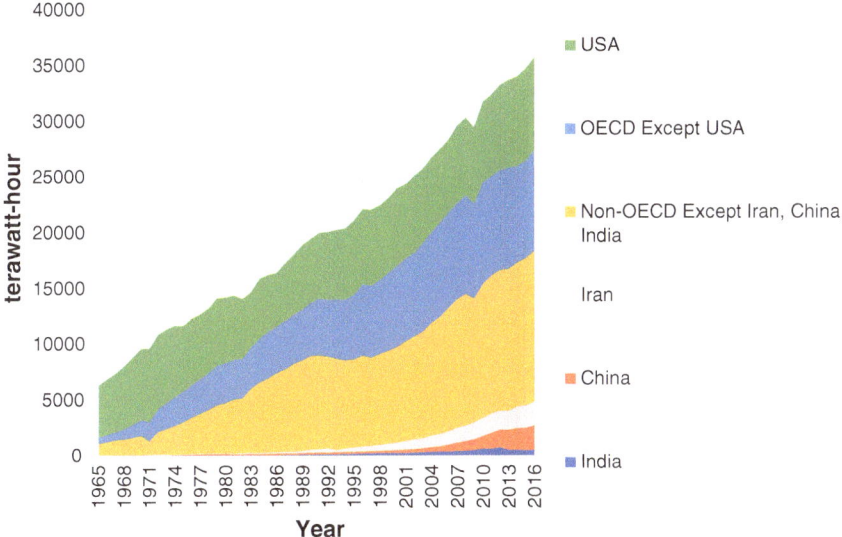

Fig. 4.11 Global natural gas consumption by country/region–a picture. *Data Source* Smil (2017), energy transitions: global and national perspectives, ABC-CLIO, LL); and BP Statistical Review 2018

Fig. 4.12 Distribution of natural gas consumption world-wide 2018. *Data Source* Adapted from BP Statistical Review 2019

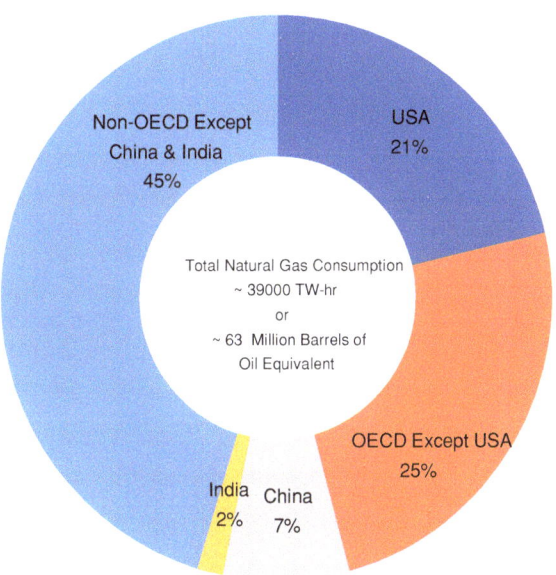

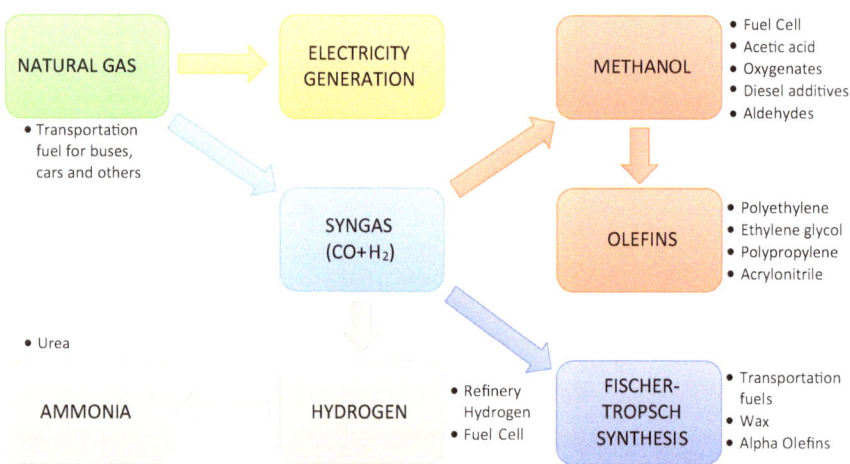

Fig. 4.13 Various uses of natural gas

to liquids because of the volumes involved—for any meaningful quantities to be confined in limited spaces, they either have to be liquefied (cold temperatures) or pressurized (high pressures), which present their own challenges.

4.5   Resource Availability—*Of the Big Three*

There has been a perpetual question: When will the primary sources of energy be exhausted? Particularly, oil? Fears of oil shortages were "confirmed" by USGS as early as 1906 when the global oil production was less than 500,000 barrels per day. In 1928, when the production of oil was around 3 million barrels of oil per day, US analyst Ludwell Denny, in his book "We Fight for Oil",[2] noted that "the domestic oil shortage and said international diplomacy had failed to secure any reliable foreign sources of oil for the United States. Fear of oil shortages would become the most important factor in international relations, even so great as to force the U.S. into war with Great Britain to secure access to oil in the Persian Gulf region" (until World War II, Great Britain controlled over 80% of the world's resources). Oil continues to be produced and consumed at a pace fast and furious than before—from less than 500,000 barrels per day to about a 100 million barrels today (2019)—to fuel Man's ever-expanding mobility and serve as a source/key ingredient of many products of modern life. Detailed discussion of oil/gas/coal resource availability/access and 'peak oil' are beyond the scope of this work. However, suffice it to say that based on available data from sources such as USGS/EIA/BP Statistical Review 2019, there

---

[2]Denny (1928).

are enough conventional/unconventional fossil fuel resources accessible per existing technology to last for hundreds of years with an annual demand of over 400,000 TW-hr of energy (OECD projections; compared to today's (2019) energy demand of about 150,000 TW-hr)!

4.6 Tailpiece

As things stand today, there is no substitute to the 'Big Three'—all natural sources of energy made *by* Earth with '*help*' from the Sun—to drive the Western way of life also being emulated as a model of development and economic growth by the rest of the world notwithstanding the calls for alternate/renewable/green energy, with all due respects. As with everything in life, there is but a price to pay in terms of energy and the environment—there is no free lunch! As an example, 1 million tons of oil equivalent has about 12 terawatt-hr of energy but produces only about 4.4 terawatt-hr of electricity—Second Law of Thermodynamics is in play: "you cannot break-even" (see Chap. 2); over 60% of the original energy is 'lost'. Environment must not be degraded; but, infrastructure development does precisely that! Burning fossil fuels emits carbon dioxide, identified as a green-house gas, but to pin the blame on it as the primary cause of global warming, and now the new term of climate-change, is too simplistic. No doubt, climate-change is real and Earth has cycled through ice-ages and warming periods over geologic time—carbon dioxide is also naturally occurring and not all of it is man-made; it is responsible for plant growth, for the oxygen on earth, it is entrapped in the oceans, in naturally occurring carbonate deposits, and the like. As we walk through the following chapters on alternate/renewable/clean energy, they will provide more food for thought and added understanding for sifting the grain from chaff. Ultimately, again, *there is no free lunch—and we all are caught between a rock and a hard place!*

References

Denny L. (1928). *We fight for oil, createspace independent publishing platform.* New York: First published Alfred A. Knopf Inc., 25 Jan 2017.
Smil, V. (2017). *Energy transitions: Global and national perspectives.*

Chapter 5
Alternate Energy—*Myth and Reality*

5.1 Laws of Energy—*Laws of Nature; Eternal and Inviolable*

A quick reminder (from Chap. 2): *Not only can you not win, you cannot breakeven either; you will always lose—there is no free lunch!*

Furthermore, what started as calls for alternate energy in the mid-1970s, metamorphosed over time into calls for renewable energy, clean energy and green energy. Interestingly, the only renewable energy is 'carbon energy'—in the form of wood (or biomass) over Man's life-span; and, fossil fuels over Earth's life-span. Solar energy is just 'there'—and so is 'Wind'; because of the Sun.

At its most basic, Man needs carbon energy to live—in the form of food that we eat; and, of course oxygen to breathe, the precursor of which is carbon dioxide.

5.2 Green—*Is Also 'Black'*

The only 'Green' energy on Earth, in the present day meaning of the term, is that received directly from the Sun in its raw, nascent and pristine form. Period. Full stop. Everything else is 'Black'. Primary Energy, discussed earlier, is of course 'black' (in the present-day definition of the word) but so is all Alternate Energy.

Sun is responsible for all life on Earth—take the Sun away and all life on Earth, as we know it, ceases to exist. In fact, Earth harnesses solar energy in its own way and makes it useful for Man—life.

Sun is responsible for the hydrologic cycle (Fig. 5.1)—clouds to form, rain and snow to fall, glaciers to form and melt, rivers to flow, water to drink, changes in seasons, wind and weather patterns.

Sun is responsible for the carbon dioxide cycle (Fig. 5.2)—plants to grow, carbonaceous rock and other material to form, including fossil fuels, and all oxygen on Earth. Carbon (CO_2) is not man-made and was on Earth before Man and will be there after

© Springer Nature Singapore Pte Ltd. 2020
R. Sharma and V. Pareek, *The World of Energy*,
https://doi.org/10.1007/978-981-15-6724-7_5

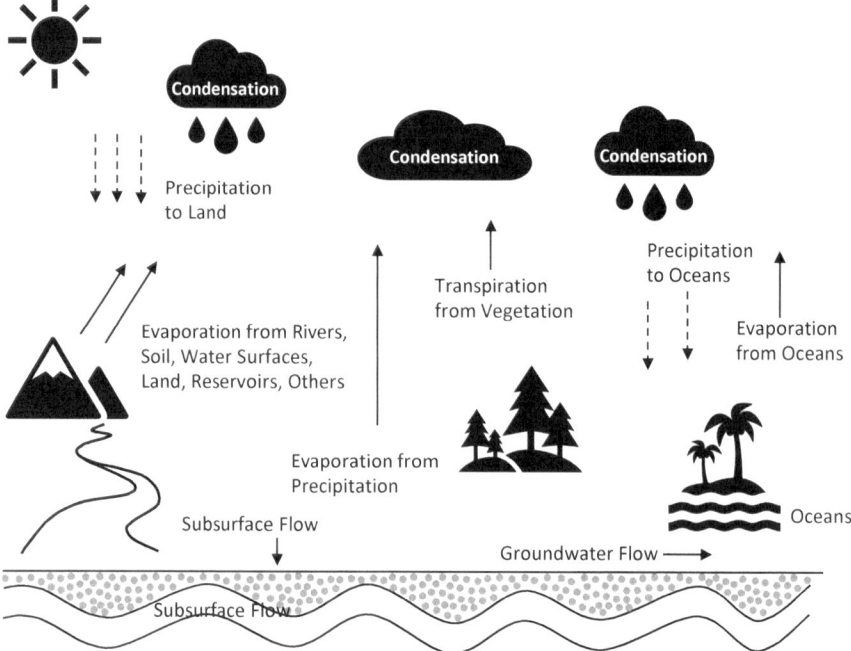

Fig. 5.1 Water cycle

Man. Nature has provided an automatic clean-up mechanism for carbon waste in the form of the CO_2 cycle (Fig. 5.2)—this is not so with nuclear waste, or electronic waste, or plastic waste, or any other waste including that caused by 'Alternate Energy' and Man's desire for ever-more!

It is the 'clean, green, energy'—solar and wind in their raw, natural state—that gave Man his start and it is this energy than Man left behind in his quest for ever more. Sun's energy on earth is available in a highly diffused form and only for part of a day at any one place—not conducive 'naturally' to provide the capacity to do work on a scale and speed required by Man for the modern comforts of life. Intervention by Man is required to harness this energy in an *instantly* useful form—and that is where the problem begins. Similarly, with Wind—wind cannot be ordered to blow, or its direction, or speed; the energy it provides is intermittent and not constant. Harnessing solar and wind energy to make them instantly useful to Man, makes them 'black'. Old, existing ('black') energy is required to manufacture mechanisms to harness solar and wind energy (in the form of electricity)—these manufacturing processes produce their own liquid and solid waste, in addition to the carbon dioxide emissions from the use of 'black' energy; all environmental pollutants.

Man has enriched himself at the expense of the Earth—not so with any other living species on Earth. *There is no free lunch!* Further, pollution once caused cannot

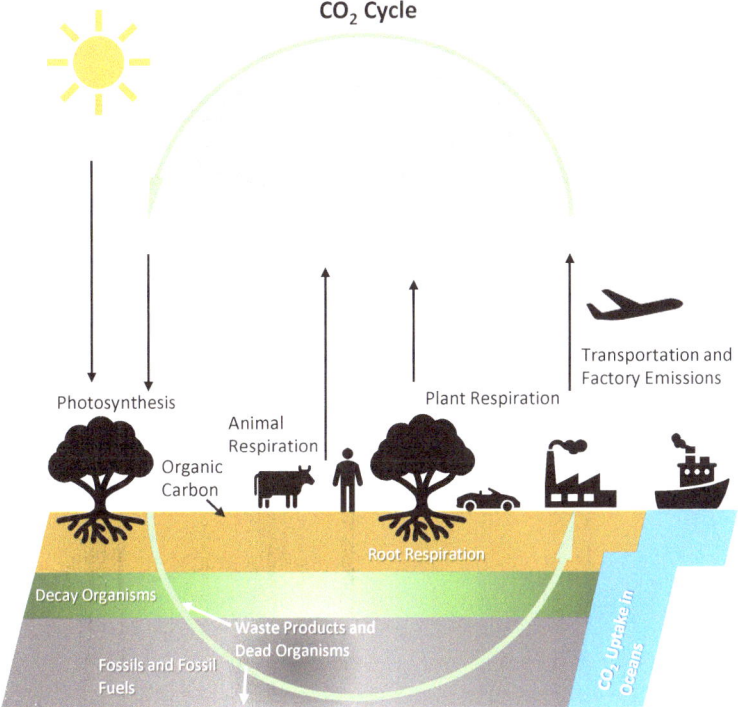

Fig. 5.2 Carbon dioxide cycle

be destroyed—only its form may or may not be changed; from gas to liquid to solid or either which way.

The following sections discuss the various alternate energy forms, as presented in Fig. 3.14, for the reader to form his/her own informed views on the matter.

5.3 Solar Energy—*Earth's Workforce*

Energy delivered by the Sun, in the form and intensity delivered, provides Earth the capacity to do its work—sustain all life and the resources required to do so! The form and intensity of the energy received by Earth is perfect for Earth to do its work—but, not so for Man to do our work at the scope, scale, and speed required for providing the modern (so-called) comforts of life. It is true that the aggregate of solar energy received by Earth in one hour, in terms of the absolute quantity (approximately 690,000 TW-h; Fig. 5.3) is many times that consumed by Man in one year for modern day living (approximately 154,000 TW-h in 2017; Fig. 3.3)—yet, Its contribution to modern day living remains at less than one per cent of the global energy consumption (Fig. 5.5). Much as solar energy is so perfect and vital for all life and resources on

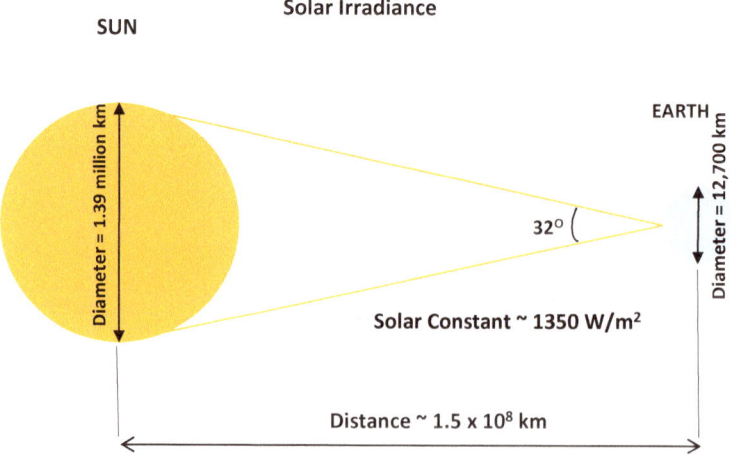

Fig. 5.3   Solar Constant

Earth, it is highly diffused received in its raw form and lacks the power punch to do the required work for delivering modern comforts of life to Man. However, Man's efforts at harnessing and harvesting this energy naturally through the ages is well known—as examples, mud houses, in use till about a hundred years ago (and still in use in some parts of the world today) would be cool on the inside during Summers and warm during Winters; earthen pots were used (and still are in some parts of the world today) to provide cool drinking water. Not only these, there are other examples of ancient architecture, particularly in the East, harnessing and harvesting solar energy for ease of living. The science and engineering behind all this are well known. Mud houses and earthen pots, ancient architectural designs of forts, palaces, and buildings, are all examples of natural, passive use of solar energy—completely 'Green'.

Man's efforts at harnessing and harvesting solar energy for producing electricity required to provide modern comforts of life is at variance with the use of solar energy in the past as explained above. In recent times, there has been a lot of talk about electricity produced from solar energy as being clean energy—this is far from being completely true! Solar collectors and photovoltaic panels needed to produce electricity from the Sun's rays require 'old energy' (carbon energy) to manufacture; not only that, they also require minerals and chemicals. Were the solar collectors and photovoltaic panels to be found 'naturally', then perhaps electricity produced from solar energy could be considered 'clean' energy. Not only old energy, minerals and chemicals, are also required to manufacture solar collectors and photovoltaic panels, the manufacturing processes to produce these equipment require substantial quantities of water and create water pollution. In fact, the pollution has just been transferred from 'air' to land and water. Furthermore, electronic waste produced as a

Table 5.1 Qualitative comparison of two major solar technologies—solar thermal and photovoltaics for electricity generation

	Solar thermal	Photovoltaics
Renewable	Yes (*)	Yes (*)
Capital costs	Very high	Very high
Operating costs	Moderate	Moderate
Costs/kW-h	More expensive than 'conventional power'	More expensive than 'conventional power'
Efficiency	~15%	~5–10% (Thin films ~ 4–5% (Lab))
Storage	*	*
Pollution	Thermal* (Air and water pollution during the manufacturing process of collectors)	Thermal (Air and water pollution during the manufacturing process of photovoltaic panels)
Environmental impact	Moderate	Large
Scale	Large—possible Small—no Integrated with 'fossil power'	Large—depends on area small—possible*

*Limitation/uncertaintities.

result of these collectors and panels has no 'automatic' clean-up mechanism, unlike the nature-provided CO_2 cycle, requiring further handling and processing. The point being that, *there is no such thing as a free lunch*! It is critical to understand that pollution once created cannot be destroyed—only its form can be changed from one state of matter to the other within the natural states of solid, liquid, or gas, with its own impact on the environment. A quick comparison of the two pathways of producing solar electricity is presented in Table 5.1. Just to re-emphasize, electricity is an energy carrier and not an energy source (Chap. 2).

Solar thermal processes have also been studied for 'passive' (radiation, convection, evaporation) residential heating and cooling vis-à-vis in the modern context in the form of a fully automated and instrumented experimental Solar House(s) designed and built on the Colorado State University Campus in Fort Collins, Colorado, USA, in the early/mid 1970s funded by NSF. Even in these experimental houses, the entire heating and cooling loads could not be met by solar energy and conventional (natural-gas fired) back-up systems were required for space and water heating. Even in these 'solar houses', use of solar energy was not really 'passive' as man-made 'collectors' were required to concentrate this energy for driving man-made heating and cooling processes—again dependent on 'old' energy, mineral and chemical resources as discussed above. Today, almost five decades later, mass adoption of these heating and cooling systems remains elusive, though small, individual, heating systems (encouraged by government subsidies) are being used in many countries.

That said, solar energy is available when it is available—cycle time for solar energy is approximately twenty per cent in a day unlike energy from fossil fuels which is available 24 × 7! Storage systems have yet to be devised to make solar energy attractive as a power source when Sun is not shining. Further, low efficiencies of 'converting' solar energy into electricity via photovoltaics require intensive use of land area plus perhaps cause 'thermal pollution'. As an example, a 1000 MW plant situated near the equator would require about 70 km^2 of land area[1]—many, many times the land area required by an equivalent capacity 'fossil-fuel' fired power plant. One can only imagine the thermal pollution ('heat island' effect) generated from a 'normal' sized 1000 MW electricity generating plant—as an example, about 100 MW of heat would be discharged to the atmosphere for a mere 10 MW electric power facility. However, small, individual systems are being currently used (with government subsidy/incentives) in various countries.

Setting all the hype aside and coming to terms with solar energy as being the answer to provide/sustain modern comforts of life, it has been the 'future' for over 150 years and is 'expected' to be so for in the foreseeable future (photoelectric effect was discovered by Edward Becquerel in 1839; first flat plate collector used in modern solar water heating was patented in 1891 in Baltimore, Maryland, with the first commercial systems developed in California by William Bailey in 1909; but, the collector market had collapsed by 1920 because of natural gas).

5.4 Wind Energy—*Solar's 'Child'; Unpredictable at Best*

Wind is air in motion. Local winds are caused by heated air above the land mass expanding and rising with the cooler air from above the water bodies gushing into fill the void, with the phenomena reversing in the evenings after sun set—this because the land mass heats more quickly by the sun's radiation than water bodies; and cools off more quickly too! Atmospheric winds, large mass of air swirling around the earth, are caused by the land mass near the equator being hotter than near the North and South Poles.

Man has tried to harness the kinetic energy of wind, energy as a result of the motion of air, through the ages. Thousands of years ago, wind was first used to sail ships. Wind mills later appeared to do other 'work' such as pump water to grind wheat. They are said to have originated in China a couple of thousand years ago (with some references claiming to be in 1500 B.C. Persia) and centuries later migrating to Europe. Domesday Book in 1086 records 6082 water or wind mills in England. In the year 1300, there were said to be 12,000 mills.[2]

---

[1] https://energy.mit.edu/wp-content/uploads/2015/05/MITEI-The-Future-of-Solar-Energy.pdf (calculation based on net power density of 15 W/m^2—page 266).
[2] Fouquet and Pearson (1998).

In the early part of Twentieth Century, people in rural United States used windmills to generate electricity before power lines began transmitting electricity from fossil-fuel fired power plants to these areas starting in the 1930s, and wind turbines were phased out. The oil embargo of the 1970s, as discussed in the Chap. 1, gave rise to calls for alternate energy (even though oil and electricity do not 'mix) and wind mills were back on the agenda for electricity production—it is close to 50 years since then and wind energy has yet to make any significant impact as discussed in an earlier chapter.

Wind energy is unpredictable and unreliable—it is available only when the wind blows; and that too only between certain speeds (8–25 mph). Neither can wind be ordered to blow; nor can the velocities be fixed. Electricity produced on windy days cannot be stored during calm periods and conventional power plants must serve as back-up—capacity factor for wind produced electricity ranges between 30 and 40%.

There are other limitations that are glossed over by proponents of wind generated electricity such as 'old energy', materials and resources required to produce wind turbines; such as land area requirements, environmental concerns of the turbines themselves; and, the true cost of wind produced electricity. As an example, just to power Houston's requirement of 2700 MW-day of electricity, the turbine-forest would impact more than half-a-million acres (900 square miles) of land[3]; whereas, a 500 MW gas-fired power plant requires about 15 acres; and, all the nuclear power plants in the United States take up less than 75,000 acres (but more on nuclear energy in a later section).

Wind generated electricity is also 'sold' as environment-friendly with the focus being only on 'emissions' but other kinds of environmental impacts are conveniently forgotten. As an example, just in Northern California's Altamont Pass, thousands of birds (including eagles, hawks, owls, and other birds of prey) are killed every year by wind turbines in violation of bird protection laws.[4] Noise pollution generated by wind turbines affect quality of life and health of people living a quarter of a mile.[5] Hundreds of deaths and other serious wind turbine related accidents, though infrequent, have also been reported over the past two decades.[6] Even for 'emissions', Courtney has argued and concluded that wind farms provide "no reduction to the need to operate conventional thermal power stations and makes little or no reduction to emissions from them".[7]

In addition to the above, wind energy is expensive. Subsidies, tax incentives, regulatory framework (requiring 'power companies' to buy wind-generated electricity) mask its true costs. A study conducted by Utah State University in 2016, examines in detail these matters and states "w(W)hen examined more closely, many claims about

[3]Landry (2018).

[4]Thelander et al. (2003).

[5]Jeffrey et al. (2013).

[6]www.caithnesswindfarms.co.uk.

[7]Courtney (2006).

wind energy are found to be indefensible."[8] In fact, wind generated electricity is many times more expensive than that generated from natural gas fired power plants.

Danish experience with wind generated electricity is held up as a beacon for the world to follow. It is instructive to know a few facts about that as well. Sharman[9] reveals, "Denmark is exporting most of its wildly fluctuating wind power to larger neighbors while finding other solutions for supply and demand at home".

Since there is no storage of electricity in any transmission and distribution system, load and demand are balanced dynamically in all power grids. Danish grids have strong inter-connections with neighboring countries but are not linked to each other and Denmark makes full use of these to balance wind power. "However, the inter-connectors were built primarily to link Norway and Sweden to Germany and, without their prior existence, it may not have been viable for west Denmark to build wind capacity on the scale it has" [12]Further, needless to say, over the past 50 years, Denmark has provide(d)s extensive subsidies and incentives to develop and maintain its wind energy systems.

We need to be realistic about what can and cannot be done with wind energy.

5.5 Biofuels—*Back to the Future; Also 'Black' as Else*

Biofuels have been the future of transportation for over a hundred-and-fifty years ever since petroleum was discovered in 1859 (in any large measure in Pennsylvania) and its use became gradually prevalent. Nicholas Otto, German inventor and father of the modern internal combustion engine (known as the Otto Cycle; 1876), first used ethanol as fuel in one of his engines in 1860. In fact, Henry Ford made his first automobile in 1896, the quadricycle, to run on ethanol. It was on the advice of Thomas Alva Edison that Ford switched to gasoline and made the Model T in 1906 as a flexi-fuel vehicle to run on both gasoline and ethanol, or a combination. By the 1920s, gasoline had become the fuel of choice and all transportation (land, sea, and air) had become 'petroleum-based'. The 1930s saw fuel ethanol gain a market in the Mid-West United States with over 2000 gas stations (petrol pumps) selling gasohol (6–12% blend of ethanol with gasoline). The War years, 1941–45, saw an explosion in the use of petroleum to drive the military machine. Even though production of ethanol increased significantly during these years, it was mainly for non-fuel war-time uses. After the War, for over three decades till the late 1970s, there was no commercial fuel ethanol available anywhere in the United States. It was the Arab Oil embargo of the early 1970s that revived an interest in alternative transportation fuels, including synthetic fuels from coal, and gasohol (now defined as at least 10% ethanol blend with gasoline) was again encouraged by government subsidies as a means of 'conserving' petroleum and lessening the dependence on Middle-East oil.

---

[8]Hansen et al. (2016).
[9]Sharman (2005).

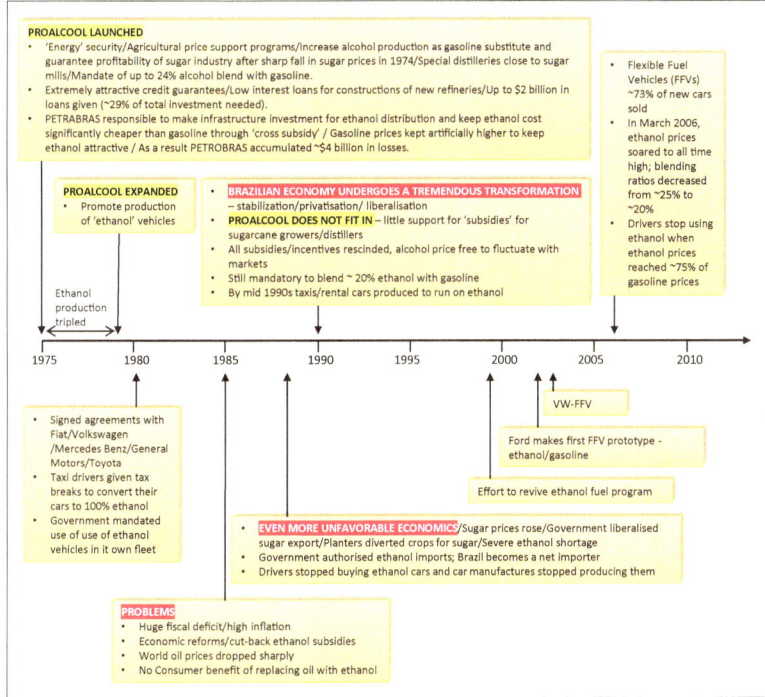

Fig. 5.4 Brazilian ethanol experience over thirty years. *Source of information* information from Xavier, 2007, the brazilian sugarcane experience, competitive enterprise institute (issue analysis), no.3 (feb.)

Over the past century, interest in ethanol has been driven mainly by economic and political factors.

In the above context, it is also interesting to view the Brazilian experience with ethanol as a transportation fuel. Figure 5.4 presents a summary of this experience over the past 40 years starting in 1975. The Brazilian experience did not turn out well. "If ethanol were truly key in displacing oil imports, the Brazilian ethanol program also shows that biofuels should not be considered a panacea for the world's energy challenges. Brazil's ethanol infrastructure model required huge taxpayer subsidies over decades before it could become viable. The ethanol program became uneconomical when petroleum prices fell in the late 1990s. Even today, during a period of high oil prices, ethanol volatile prices have not freed Brazilians from losing money on the E20 blend mandated by their government".[10] Further, overlooked has been the fact that oil self-sufficiency had been a long-term goal of the Brazilian government which was achieved in 2006 with Brazil now a net exporter of oil.

---

[10] Xavier (2007).

Lately, the focus is back on biofuels—this time ostensibly because of environmental concerns. They are now being propagated as being more environment friendly and less polluting than gasoline in terms of carbon (Carbon Dioxide) emissions. Let us for a moment focus on this and dispel some myths. Energy content of ethanol (~20 MJ/l) is about two-thirds that of gasoline (~32 MJ/l) requiring one-and-a-half times more ethanol than gasoline to travel the same number of miles Carbon (CO_2) emissions from burning a gallon of gasoline are about 19. 5 lb (~8.8 kg); and, about 12.72 lb (~5.8 kg) from burning a gallon of ethanol. However, since it takes one-and-a-half gallons of ethanol to travel the same number of miles as a gallon of gasoline, carbon emissions from burning ethanol on a per mile basis are about the same as from gasoline. So it is, for biodiesel in terms of carbon emissions. Hence, there is no real advantage of biofuels over gasoline. In fact, gasoline packs a greater power punch on a per unit volume basis and requires a smaller fuel tank in a car than for ethanol to travel the same distance. An argument is made that biofuels are renewable—true; in the sense that crops can be planted, grown, harvested, and processed into biofuels several times during a Man's life-time. So are fossil fuels over geologic time—made by Earth from plant material with help from the Sun, ready for use. Biofuels are also said to tend towards being carbon-neutral—biomass requires carbon dioxide to grow before emitting carbon dioxide when used to provide energy; this debate is far from settled depending upon system boundaries considered for analysis. Perhaps, in a broader sense, fossil fuels are carbon-neutral as well—after all, their origin is also plant material sequestered carbon dioxide over geologic time! Furthermore, not only do biofuels take land away from growing food, the jury is still out on whether they are energy positive—giving off more energy on burning than that required to plant, grow, harvest, and process crops into biofuels. Even so, they are nowhere near as energy positive as fossil fuels. Figure 5.5 presents energy content of some liquid fuels as comparison.

After all, biofuels are also carbon energy. There are no quick and easy solutions.

5.6   Hydrogen—*The Reality*

Hydrogen is the lightest and most abundant element in the Universe. About ninety percent of the Universe mass is hydrogen. The main essence of stars is hydrogen including our Sun which is mostly hydrogen and part helium. That be it, free terrestrial hydrogen does not exist and must be made from substances containing hydrogen such as water and hydrocarbons. We know how to extract hydrogen from these sources even on a commercial scale for specific applications with added value other than for supplying energy. To use hydrogen to drive the world as we see it today is fraught with many consequences. Figure 5.6 presents various pathways under different stages of development for production of hydrogen. Regardless of the pathway, *Hydrogen is not an energy source but an energy sink*—energy content of extracted hydrogen is much less than that required for it to be produced.

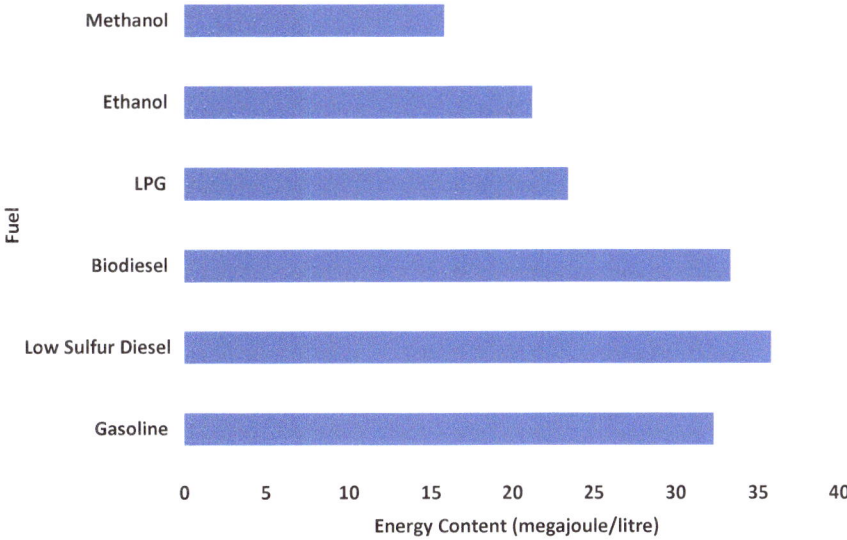

Fig. 5.5 Comparison of energy content of various liquid fuels. Data from www.afdc.energy.gov

As hydrogen is the lightest element and exists as a di-atomic gas at ambient conditions, special handling and storage are required. Even though the energy density of hydrogen on a mass basis is high (~52,000 Btu/lb or ~120 MJ/kg), on a volume basis it is very low (~275 Btu/ft^3 at 60 F and atmospheric pressure or ~10 MJ/Nm^3). The energy content of one kilogram of hydrogen is approximately the same as that of a gallon of gasoline. To store a gallon of gasoline (a little less than 3 kg), requires about a 4-L vessel at ambient conditions (atmospheric pressure and temperature); whereas, to store 1 kg of hydrogen at ambient conditions requires about a 12,000-L capacity vessel. The volumetric energy density of hydrogen can be increased by storing it at high pressures or low temperatures. To store 1 kg of hydrogen in a 4-L capacity vessel would require very high pressures of the order of 3000 psi (~200 atm.). Even at higher pressures than these, the volumetric energy density of hydrogen is no match for a 40-L capacity automobile tank holding 10 gallons of gasoline. The other option of course to increase the volumetric energy density of hydrogen is to liquefy it which requires cryogenic temperatures. Hydrogen can only be liquefied at about −253 °C or 20 K at atmospheric pressure, and its density is about 70.8 g/L requiring about a 140-L capacity tank to hold 10 kg of liquid hydrogen for energy equivalency with 10 gallons of gasoline. High pressures and cryogenic temperatures have their own special energy, material of construction, design, handling, storage, distribution, and operating requirements in addition to specific safety considerations associated with hydrogen. Also, it may be surprising to many, a gallon of liquid hydrogen is less hydrogen than that contained in a gallon of gasoline. Proponents of hydrogen economy have been raising the bogey for over 15 years at least of disruption in oil supplies, peak oil, and environmental concerns forgetting that hydrogen is not an

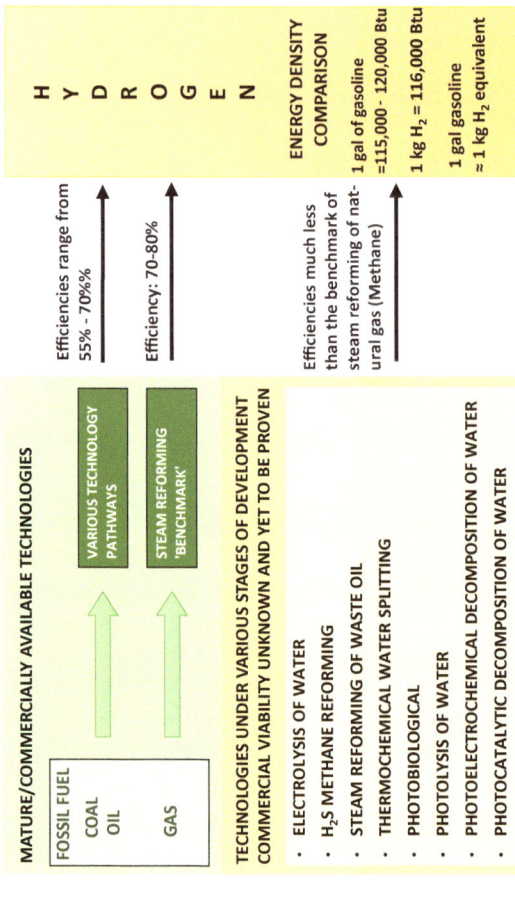

Fig. 5.6 Hydrogen production pathway. *Source of Information* Ali T Raissi and David Block, 'Hydrogen: automotive fuel of the future', IEEE power & energy, vol 2, 6, Nov-Dec 2004

energy source but an energy carrier and an energy sink and still has to be made—each of these is highly debatable and beyond the scope of this book.

Interestingly, Arno Penzias, 1978 Nobel Laureate in Physics said in an interview in 2005 "Yes, I have been and I'm still enormously skeptical about most of the solutions for alternatives. You have to do end-to-end accounting and a lot of the people who do this stuff are what I would call high verbal, low math. People say when hydrogen burns it produces only water. Did you know that hydrogen is a greenhouse gas? Nobody thinks about it, right? It's my Pit Bull. He's a sweet dog, and unless he's threatened by you, he's not going to bite you".[11] In fact, the whole interview is worth reading.

5.7 Geothermal Energy—*From Earth's Womb*

Geothermal is one of the oldest sources of energy known to mankind. Based on archaeological evidence, the indigenous communities of North America started using geothermal energy from geysers at least 10,000 years ago. All early usages of geothermal energy were direct and related to basic functions, i.e., the heat of the geothermal stream (springs, geysers) was used either to stay warm, take bath or cooking food. First community-level indoor heating systems using geothermal energy were installed in the French city of Chaudes-Aigues in fourteenth century, which are still operational. First geothermal power plant to produce electricity was built in the early twentieth century at Larderello in Italy. However, it took nearly 50 years before another power plant was built (this time in New Zealand).

The source of geothermal power is the energy that earth could not get rid of as it cooled down after its formation; the energy being a result of underground radioactive decays. At a depth of nearly 6500 km, earth's innermost part, is a solid ball of fire with temperatures as high as 5000 °C. Typically, the temperature increases by 3°C every 100 meters of depth. Thus, at a depth of about 7 km one could expect a temperature of about 200 °C—enough to power a steam engine to produce electricity. Unlike wind, tidal and solar energy, geothermal energy is available uninterruptedly. Therefore, theoretically, we have an enormous source of energy right under our feet—enough to power the civilization for thousands of years. However, practically there are several scientific and technical challenges. For example, most of the geothermal energy is either inaccessible or available at low temperatures (30–50 °C)—heat at such temperatures cannot do any work or produce electricity. Such heat is only useful for local community heating systems; and that too only in colder places. For example, a 50°C geothermal source can only be of recreational use in tropical regions; but, it could be a boon for arctic countries such as Iceland, where 90% of heating requirements are met using geothermal energy. Despite higher capital costs, in wealthy countries such as Australia, the low temperature geothermal energy is becoming popular for swimming pool heating. Since it involves no additional step

---

[11] https://www.cnet.com/news/from-the-big-bang-to-big-bucks/.

between production and consumption, direct heating is also the most efficient use of geothermal energy.

However, in order to transport the energy from the source to the end user, geothermal energy must be converted to electricity. For this to be economically feasible, the available heat must be at 200 °C minimum—higher the temperature, greater the efficiency. Such geothermal sources, at practical depths, are mostly confined to regions of volcanic activity. Not surprisingly, some of the largest producers of the electricity using geothermal energy are in regions located along the Ring of Fire in the Pacific Ocean. The list of countries located on this Ring includes, New Zealand, the Philippines, Indonesia and United States. Even so, the net contribution of geothermal energy globally is miniscule. For example, with nearly 30% of world's production, the US is the largest producer of geothermal energy, and yet it only constitutes less than 0.35% of the US energy needs (which is negligible; Fig. 1.2).

High capital costs and lack of adequate research on extraction of heat are often proposed as the primary reasons for the low uptake of geothermal energy. However, being location specific is indeed the primary reason for the lack of uptake—it is not available everywhere near the earth surface. The prospects of deeper geothermal wells (essentially deep holes) have not progressed due to seismic effects they can cause. For example, in 2006, Switzerland had to abandon its plans for a new geothermal power station after a series of earthquakes were triggered during the drilling process. Even for shallow geothermal wells, there is a possibility of groundwater contamination. Geothermal wells may also produce hydrogen sulfide, a gas that smells like rotten eggs, which even at low concentrations (as low as 2–5 parts per million) is toxic if inhaled over a long time.[12] Some wells may also produce unwanted geothermal fluids containing toxic substances requiring costly disposal and waste management. Finally, all geothermal sites cool down progressively; as an example, New Zealand's Wairakei geothermal reservoir has cooled down by 20 °C over the past 60 years; and new wells may have to be drilled to keep the operation running, thus requiring new capital investments progressively. These drawbacks have limited the use of geothermal energy to less than 0.06% (negligible) of the total global energy consumption (Fig. 3.5).

5.8 Nuclear Energy—*Ever Hazardous*

In effect, it started with Albert Einstein's special theory of relativity given to the world in 1905 and his most famous equation, $E = mc^2$, establishing that energy and mass are interchangeable; m is the mass and c is the speed of light (~186,000 miles per second or about 300,000 km per second), This equation set the stage for nuclear energy—nuclear power and nuclear bombs.

Leo Szilard, a collaborator of Einstein, in 1933 discovered the nuclear chain reaction that could release the energy locked up in atoms. 1933 was also the year that

---

[12]https://www.osha.gov/SLTC/hydrogensulfide/hazards.html.

Adolf Hitler became the Chancellor of Germany. By 1939, Szilard was convinced that German scientists could be on the path of developing an atom bomb. He persuaded Einstein, who had earlier migrated to the United States, to warm President Franklin Delano Roosevelt prompting Einstein to write his famous letter to the President and advising him to fund atomic energy research. President Roosevelt took this letter seriously and appointed an Advisory Committee which met for the first time on October 21, 1939, just a few days after Hitler's invasion of Poland and a couple of months after the 'letter'. This Committee was the pre-cursor of the Manhattan Project[13] which led to the development of the atom bomb and the bombing of Hiroshima and Nagasaki in August of 1945. United States entered the Second War after the attack on Pearl Harbor by Japan in December of 1941. Enrico Fermi led a group of scientists in Chicago to successfully initiate the first self-sustaining nuclear chain reaction on December 2, 1942. J. Robert Oppenheimer was the Head of the Los Alamos National Lab in New Mexico where the 'bomb' was finally developed and tested under the Manhattan Project and is called the 'Father of the Atomic Bomb'. Of course, a lot of coordinated effort had gone on since 1939 at various academic institutions and labs in the United States before culminating in the Manhattan Project.

First use of nuclear energy had been for weapons and war! Death, destruction, and after effects of the bombing of Hiroshima and Nagasaki are all too well-known.

Following the War, even though substantial effort continued on both sides of the 'Iron Curtain' (the then Soviet Union and the West) for the development of nuclear weapons, scientists realized the potential for harnessing the tremendous amount of heat generated during nuclear reactions for generating electricity and various other applications. The first instance of generating electricity from nuclear energy was at the Argonne National Lab in Idaho, USA, on December 20, 1951, in a small experimental reactor. President Dwight D. Eisenhower's 'Atoms for Peace' speech at the UN General Assembly in December of 1953 paved the way for significant research effort in the United States directed at using nuclear energy for electricity generation and other civilian applications.

Parallel developments were happening in other parts pf the world—particularly, the Soviet Union and England. The first atomic power station (5 MW-electricity or 30 MW-thermal capacity) began operation in Obninsk in June 1954, supplying electricity to the grid. The first full-scale power station (268 MW-thermal or 35 MW-electricity capacity) came on-line at Calder Hall, in England, in October, 1956. It was later revealed,[14] however, that this plant was primarily set up to produce plutonium for UK's nuclear weapon's program and electricity was a 'by-product'

Almost seven decades after the first experimental generation of electricity from nuclear energy in 1951 in Idaho, USA, nuclear power provides only about 2% of world's energy needs (Fig. 3.5). There are several major issues related to nuclear power that need to be considered seriously.

---

[13]https://www.osti.gov/opennet/manhattan-project-history/Events/1939-1942/uranium_research.htm.

[14]https://www.ice.org.uk/what-is-civil-engineering/what-do-civil-engineers-do/calder-hall-nuclear-power-station.

Nuclear power plants are highly capital intensive with costs many times that of conventional fossil-fuel fired power plants ranging between $6 and $8 billion for a 1100 MW[15] plant compared to less than $1 billion for a similar size coal fired plant. In addition to higher operating life-time costs compared to conventional power plants, decommissioning of nuclear power plants is a time-consuming and an expensive process costing billions of dollars. As an example, the Three Mile Island nuclear power plant in Pennsylvania, USA,, which came on-line in 1974, shut down in the second half of 2019 and will take 60 years for it to be fully decommissioned. The Calder Hall plant which started operation in 1956, shut down in 2003 after 47 years of service and is expected to be fully decommissioned by 2120 at a cost of more than £70 billion.

Three major accidents, the 1979 Three Mile Island (USA), the 1986 Chernobyl (then USSR), and the 2010 Fukushima (Japan), in recent memory have led to serious concerns about safety, environmental and health effects. There have also been other accidents at fuel cycle facilities in USA, Russia and Japan. Safe and secure transportation of nuclear materials and security threats from terrorist attacks are also major concerns.

Nuclear proliferation as a result of misuse of commercial and associated facilities to acquire weapons capability is also a cause of major concern. This is because of the dual use of nuclear materials—both for power and weapons.

Radioactive waste—uranium mill tailings, spent reactor fuel, and other waste—can remain dangerous to human health for thousands of years. Handling and disposal of radioactive waste continues to be one of the biggest challenges and remains unresolved. Unlike Nature's automatic clean-up mechanism for 'carbon waste' in the form of the $CO2$ cycle, there is no such mechanism for nuclear waste by Nature.

Nuclear energy may be carbon-free in the long run—but is ever hazardous; and, certainly not clean. If, as and when, more nuclear power plants are built around the world, the above challenges are only bound to grow.

Solution to a problem often gives rise to a different set of problems. May be different, but certainly there is always a price to be paid.

References

Courtney, R.S. (2006). *Wind farms provide negligible useful electricity.* Center for Science and Public Policy—Washington, D.C., March 2006.
Duffie, J. A., & Beckman, W. A. (1974). *Solar energy thermal processes.* New York: Wiley.
Fouquet, R., & Pearson, P. (1998). A thousand years of energy use in the united kingdom. *The Energy Journal, 19,* 4.
Hansen, M.E., Simmons, R.T., & Yonk, R.M. (2016). The unseen costs of wind-generated electricity. Institute of Political Economy, Utah State University.
Jeffrey, R. D., Krogh, C., & Horner, B. (2013). Adverse health effects of industrial wind turbines. *Can Fam Physician, 59*(5), 473–475.

---

[15]https://www.synapse-energy.com/sites/default/files/SynapsePaper.2008-07.0.Nuclear-Plant-Construction-Costs.A0022_0.pdf.

Landry, J. (2018). *Keynote—America First Energy Conference 2018*, Heartland Institute, 14 Aug 2018.

Raissi, A. T., & Block, D. (2004). Hydrogen: Automotive fuel of the future. *IEEE Power & Energy, 2,* 6.

Sharma, H. (2005, May). Why wind power works for Denmark. In Proceedings of ICE, Civil Engineering (vol. 158, pp. 66–72, Paper 13663).

Thelander, C.G., Smallwood, K.S., & Rugge, L. (2003). *Bird risk behaviors and fatalities at the Altamont Pass wind resource area—March 1998–December 2000'* National Renewable Energy Laboratory Report SR-500-33829, December 2003.

Xavier, M.R., (2007). *The Brazilian sugarcane experience.* Competitive Enterprise Institute (Issue Analysis), no. 3 (Feb.).

Chapter 6
The Fossil Fuel Bugbear of Climate Change—*A Perspective*

6.1 Climate Change—*Facts*

Climate change is real. Whether it is here and now and man-made, is debatable. Climate change could be natural and normal.

6.2 What Started it All—*The Present Day Claim About Climate Change*

Essentially, the 1988 Hansen, et al., paper[1] was what started *it* all; the present-day discussion of man-made global warming and climate change—even though limitations of the model have been explicitly mentioned and the time-span covered is a miniscule fraction of geologic time.

6.3 A Peek into the Past—*And, the Present*

To label climate change as a man-made phenomenon and to pin it on fossil fuel driven carbon dioxide emissions is too simplistic. Earth has cycled through warming periods and ice ages over geologic time (hundreds of millions of years) with rising and receding sea levels submerging land and forming deserts. Therefore, to base our conclusions that carbon dioxide emissions are responsible for global warming and

---

[1] Hansen et al. (1988).

The original version of this chapter was revised: Error in Figure 6.1 has been corrected. The updated version of this chapter can be found at https://doi.org/10.1007/978-981-15-6724-7_8

© Springer Nature Singapore Pte Ltd. 2020, corrected publication 2021
R. Sharma and V. Pareek, *The World of Energy*, https://doi.org/10.1007/978-981-15-6724-7_6

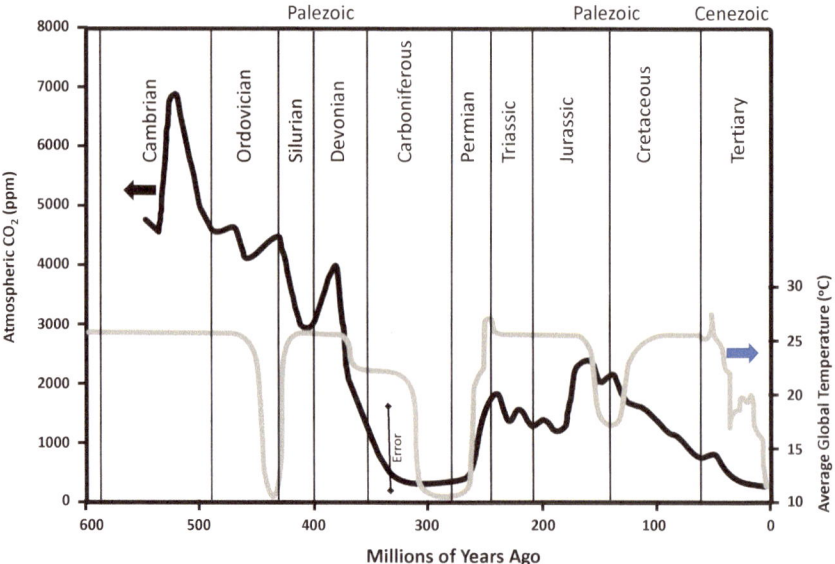

Late Carboniferous to Early Permian time (315 mya -- 270 mya) is the only time period in the last 600 million years when both atmospheric CO2 and temperatures were as low as they are today (Quaternary Period). (mya (millions of years ago))

Fig. 6.1 Global temperature and atmospheric CO_2 over geologic time. Redrawn from https://www. geocraft.com/WVFossils/Carboniferous_climate.html

climate change on a couple of hundred years (or less) of recorded data is essentially analyzing things on a 'man-time-scale' and extrapolating to 'earth-time-scale'. IPCC's conclusions about global warming, climate change, sea level rises are well known. It is not the intent here to discuss or critique IPCC's (Inter-Governmental Panel on Climate Change) findings as that is beyond the scope of this book but to make the reader aware that there is a counter-view. we wish to note that science is not by consensus but by evidence, and predictions are only predictions at best depending upon modelling, data used, and assumptions made. IPCC in Chap. 7 states, "because we do not understand the reasons for these past warming events it is not yet possible to attribute a specific proportion of the recent, smaller, warming to an increase of greenhouse gases".[2]

Figure 6.1 tracks the atmospheric carbon dioxide concentrations and temperatures over geologic time. It is interesting to note that CO_2 in the Earth's atmosphere has been much higher in the past than it is today; and, there does not seem to be much correlation between CO_2 concentration and Earth's temperature. "To the consternation of global warming proponents, the Late Ordovician Period was also an **Ice Age** while at the same time CO_2 concentrations then were nearly 12 times higher than today—**4400 ppm**. According to greenhouse theory, Earth should have been

---

[2] https://www.ipcc.ch/site/assets/uploads/2018/03/ipcc_far_wg_I_chapter_07-1.pdf.

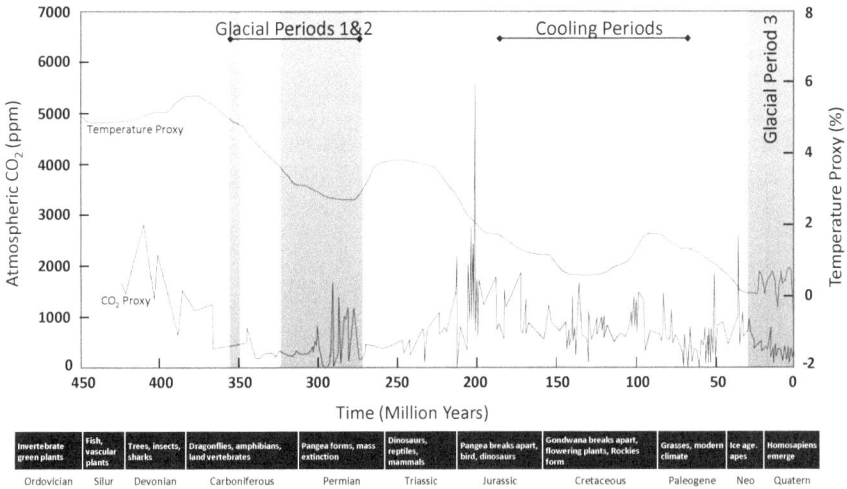

Fig. 6.2 Atmospheric CO_2 concentration over geologic time and evolutionary milestones. Redrawn from https://www.geocraft.com/WVFossils/Carboniferous_climate.html

exceedingly hot. Instead, global temperatures were no warmer than today. Clearly, other factors besides atmospheric carbon influence earth temperatures and global warming."[3] It is interesting to note that in the Paleozoic era during the Ordovician period (~450 million years ago) even with CO_2 concentration of 4000 ppm, the Earth was as 'cool' as it is today. Also, during the Carboniferous and part-Permian periods (~300 million years ago) CO_2 concentrations and Earth temperatures were as low as they are today. During the Cambrian period of the Paleozoic era (~600 million years ago), CO_2 concentration in the atmosphere was around 7000 ppm compared to around 400 ppm today—about 18 times higher!

Other studies have also concluded that CO_2 concentrations in the atmosphere may not have any effect on global climate. For example, W. Jackson Davis in the paper titled "The Relationship between Atmospheric Carbon Dioxide Concentration and Global Temperature for the Last 425 Million Years" states "this study demonstrates that changes in atmospheric CO_2 concentration did not cause temperature change in the ancient climate."[4] Came et al., in their article titled "Coupling of surface temperatures and atmospheric CO_2 concentrations during the Paleozoic era" concluded that "global climate may be independent of variations in atmospheric carbon dioxide concentration."[5]

Figure 6.2 shows evolutionary milestones as it tracks the atmospheric CO_2 concentrations over 450 million years. Carbon dioxide was *here* before Man, and shall be *there* ever after. Nature has provided an automatic clean-up mechanism (if we would

---

[3] https://www.geocraft.com/WVFossils/Carboniferous_climate.html

[4] Davis (2017).

[5] Came et al. (2007).

like to call it that) for carbon dioxide by way of the CO2 cycle as discussed in brief earlier. *Our folly seems to be to imagine 'Future-Earth' to be just like what it is today—it certainly was not in the past what it is today.* An excellent reference is how it all supposedly came into being is Neil DeGrasse Tyson's 'Astrophysics for People in a Hurry'.[6]

6.4   The Essentials—*For All Life!*

Carbon dioxide and water are essential for all life on Earth—including carbon energy (food) that we eat!

Quoting T. Huxley: "The improver of natural knowledge absolutely refuses to acknowledge authority, as such. For him, skepticism is the highest of duties; blind faith the one unpardonable sin."

We continue to learn there are no quick and easy solutions!
There is a price to be paid.

References

Came, R. E., Eller, J. M., Veizer, J., Azmy, K., Brand, U., & Weldman, C. R. (2007). Coupling of surface temperatures and atmospheric CO_2 concentrations during the Paleozoic era. *Nature, 449,* 198–201.

Davis, W. J. (2017). The relationship between atmospheric carbon dioxide concentration and global temperature for the last 425 million years. *Climate, 5,* 76. https://doi.org/10.3390/cli5040076. www.mdpi.com/journal/climate4.

Hansen, J., Fung, I., Lacis, A., Rind, D., Lebedeff, S., Ruedy, R., et al. (1988). Global climate change as forecast by goddard institute of space studies three-dimensional model. *Journal of Geophysics and Research, 93*(8), 9341–9364.

Tyson, N. D. (2017). *Astrophysics for people in a hurry.* New York: W. W. Norton & Company.

---

[6] Tyson (2017).

Chapter 7
Wrapping It All Up—*What Next?*

7.1 Something to Think About

Energy drives life. All life. In all its manifestations. Life requires work. Energy is the capacity to do work. Power is the rate of doing work—how fast. Invention of the steam engine in the Eighteenth Century, fueled by coal enabled Man to do work at a speed and scale never seen before. Industrial revolution had begun and the seeds of modern-day life sown. Coal as the energy source packed a power punch that allowed the steam engine to work at the speed and scale that it did, spawning applications in manufacturing (factories) and transportation (land and sea). Over a hundred years after the invention of the steam engine, came the invention of the internal combustion engine. Discovery of oil, the first automobile, inventions of the electric light bulb and the aero-plane, all within a span of about fifty years starting in the latter half of the Nineteenth Century, saw an explosion in our capacity to do work (and play). This explosion in capacity to do work, combined with Man's ingenuity, has provided modern day living. All this, on the back of cheap, reliable, abundant energy—coal, oil and natural gas, termed fossil fuels. Fossil fuels are sources of energy, gifted by nature—made by Earth on its own time with help from the Sun. They are not man-made, unlike electricity, for example—and, electricity is not an energy source but an energy carrier. Fossil fuels are sequestered carbon by way of the 'carbon cycle'—and it is carbon energy that drives life.

The competitive edge of fossil fuels, over other alternate forms of energy, stems from the fact that not only are they sources of energy gifted by Nature, packing a power punch at the glow of a spark, but also fungible (directly substitutable for each other); and, are sources of many other useful chemicals and products required for modern living. "While renewable resources gave us our start, renewables are also what we left behind."[1]

Lately, there have been strong suggestions that electric vehicles could be the answer to CO_2 emissions from internal combustion engine powered vehicles.

---

[1] Darmstadter et al. (1983).

© Springer Nature Singapore Pte Ltd. 2020
R. Sharma and V. Pareek, *The World of Energy*,
https://doi.org/10.1007/978-981-15-6724-7_7

This is just shifting system boundaries and a classic example of NIMBY—Not In My Back Yard! Electric vehicles still must be charged every so often from electricity produced elsewhere by whatever means with their own environmental impacts. Further, electric vehicles require Lithium-ion batteries which can weigh up to 100 kg for a 300-mile range (gasoline tank equivalent). About 10% component of this (10 kg) is Cobalt, which still must be mined and refined; and, most of the rest is Lithium Oxide, which must be extracted from Lithium Carbonate. These processes, and end-of-life disposal of batteries, (will) have their own environmental and social impacts. End to end accounting is a must - there are no quick and easy solutions!

7.2   What Next?

As things stand today, there is no replacement for fossil fuels vis-a-vis the life we lead. They will continue to dominate the energy mix in the foreseeable future, at least over the next many decades; perhaps even hundreds of years. But, then, who knows?!

It is almost impossible to predict the future, except to say that energy will continue to play a pivotal role in our lives. Watt, Otto, Diesel, Edison, Einstein, Ford, Rockefeller, Oppenheimer, and many others, could not have predicted the future then about what the world is today. Even fifty years ago, no one could have predicted the internet and its impact on the world today. As we write this, no one could have predicted COVID-19 a few months ago and its impact on the world.

Figure 7.1 presents a graphic about a possible future? Perhaps, five million years from now "Men (will be) extinct?". Who knows?!

7.3   The Only Truth About Energy

All energy, alternate energy included, is governed by, and must obey, certain laws of Nature, called Laws of Thermodynamics, which cannot be violated.

There is a cost to everything, and a price must be paid.

There is no free lunch – caught between a rock and a hard place!

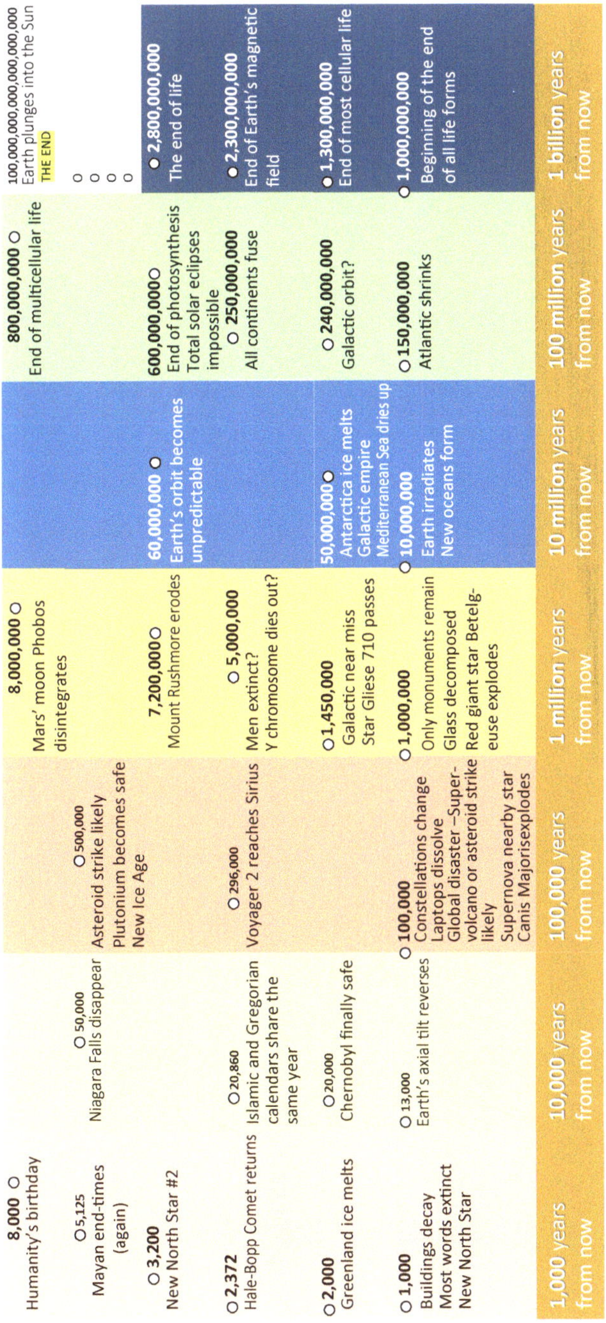

1,000 years from now
- 8,000 ○ Humanity's birthday
- ○ 5,125 Mayan end-times (again)
- 3,200 New North Star #2
- ○ 2,372 Hale-Bopp Comet returns
- ○ 2,000 Greenland ice melts
- ○ 1,000 Buildings decay / Most words extinct / New North Star

10,000 years from now
- ○ 50,000 Niagara Falls disappear
- ○ 20,860 Islamic and Gregorian calendars share the same year
- ○ 20,000 Chernobyl finally safe
- ○ 13,000 Earth's axial tilt reverses

100,000 years from now
- ○ 500,000 Asteroid strike likely / Plutonium becomes safe / New Ice Age
- ○ 296,000 Voyager 2 reaches Sirius
- ○ 100,000 Constellations change / Laptops dissolve / Global disaster —Super-volcano or asteroid strike likely / Supernova nearby star Canis Majorisexplodes

1 million years from now
- 8,000,000 ○ Mars' moon Phobos disintegrates
- 7,200,000 ○ Mount Rushmore erodes
- ○ 5,000,000 Men extinct? Y chromosome dies out?
- ○ 1,450,000 Galactic near miss / Star Gliese 710 passes
- ○ 1,000,000 Only monuments remain / Glass decomposed / Red giant star Betelgeuse explodes

10 million years from now
- 60,000,000 ● Earth's orbit becomes unpredictable
- 50,000,000 ● Antarctica ice melts / Galactic empire / Mediterranean Sea dries up
- ○ 10,000,000 Earth irradiates / New oceans form

100 million years from now
- 800,000,000 ○ End of multicellular life
- 600,000,000 ○ End of photosynthesis / Total solar eclipses impossible
- ○ 250,000,000 All continents fuse
- ○ 240,000,000 Galactic orbit?
- ○ 150,000,000 Atlantic shrinks

1 billion years from now
- 100,000,000,000,000,000,000 Earth plunges into the Sun / **THE END**
- ● 2,800,000,000 The end of life
- ● 2,300,000,000 End of Earth's magnetic field
- ● 1,300,000,000 End of most cellular life
- ● 1,000,000,000 Beginning of the end of all life forms

Fig. 7.1 Who knows? *Source of Information* The Times of India, February 14, 2014; and www.bbc.com

Reference

Darmstadter, J., et al. (1983). *Energy today and tomorrow*. Englewood Cliffs, NJ: Prentice-Hall.

Correction to: The Fossil Fuel Bugbear of Climate Change—*A Perspective*

Correction to:
Chapter 6 in: R. Sharma and V. Pareek, *The World*
of Energy, **https://doi.org/10.1007/978-981-15-6724-7_6**

The original version of this chapter was revised: Following correction has been incorporated in Figure 6.1. All points on horizontal axis scale in Figure 6.1 were all erroneously shown as '600'. They should be read as 600, 500, 400, 300, 200, 100, 0.

The updated version of this chapter can be found at

https://doi.org/10.1007/978-981-15-6724-7_6

© Springer Nature Singapore Pte Ltd. 2021
R. Sharma and V. Pareek, *The World of Energy,*
https://doi.org/10.1007/978-981-15-6724-7_8

Postface

Contrary to what the reader might surmise, it is not, and has not been, our intention to promote or propagate the use of one kind of energy over another—particularly fossil fuels. We have tried to present the facts as we see them, and the reasons why fossil fuels are such an integral part of our lives driving the modern world with the required power punch (rate of doing work).

Under no circumstances can the Laws of Nature governing all 'Energy', called the Laws of Thermodynamics, be violated. There is a cost associated with, and a price to be paid, in terms of energy for doing work—*there is no free lunch*!

Renewables gave Man his start and renewables are what Man left behind,[1] mainly because of their lack of power punch of fossil fuels and the control that they provide. Fossil fuels are also *renewables* made by Earth on its own time (geologic time), though not on Man-time, with help from the Sun. Fossil fuels are sequestered carbon by plants from CO_2—carbon dioxide was on Earth before Man and shall be there ever after. It is carbon energy that drives all life.

The emphasis today seems to be on so-called 'Green Energy'—but 'green' is not so green and is also 'black' as discussed, and does not have the competitive edge of fossil fuels. The thrust, in our view, should be on cleaner fossil fuel technologies and improving efficiency of use. This could easily give rise to the Jevon's paradox, which was also observed during the early 1980s when energy conservation and energy efficiency measures increased the overall consumption of energy. A classic example of Jevon's paradox is the increase in coal consumption in England after improvements in the steam engine efficiencies.

In ingenuity of Man we have the utmost confidence. Even though we do not foresee any substitute or alternative to fossil fuels over the next several decades at least, it is almost impossible to predict technology advances or technical changes that will drive our future. Professor Solow, Nobel Prize in Economics, 1987, said about 'technical change': "I am using the phrase 'technical change' as a shorthand expression for *any kind of shift* in the production function. Thus slowdowns, speedups, improvements

---

[1] Darmstadter et al. (1983).

© Springer Nature Singapore Pte Ltd. 2020
R. Sharma and V. Pareek, *The World of Energy*,
https://doi.org/10.1007/978-981-15-6724-7

in the education of the labor force, and all sorts of things will appear as "technical change".[2] *We continue to live and learn.*

Necessity may dictate use of renewables to produce mainly one product—electricity, an energy carrier and not an energy source. However, it must be kept in mind that cry for renewables changed from being an alternative to oil (oil and electricity do not 'mix'), to carbon dioxide emissions and global warming, and now to climate change over the past four decades. Regardless, an end-to-end accounting of energy and environment is a must before any renewables can be considered as options, or alternatives, on a mass scale, to replace fossil fuels.

Raj Sharma
Vishnu Pareek

References

Darmstadter, J. et al. (1983). *Energy today and tomorrow.* Englewood Cliffs, NJ: Prentice-Hall.

Solow, R. M. (1957). Technical change and the aggregate production function. *The Review of Economics and Statistics, 39*(3), 312–320. MIT Press, Article Stable http://www.jstor.org/stable/192607.

---

[2]Solow (1957).